Birkhäuser

Trends in Mathematics

Research Perspectives Ghent Analysis and PDE Center

Volume 12

Research Perspectives Ghent Analysis and PDE Center is a book series devoted to the publication of extended abstracts of seminars, conferences, workshops, and other scientific events related to the Ghent Analysis and PDE Center. The extended abstracts are published in the subseries Research Perspectives Ghent Analysis and PDE Center within the book series Trends in Mathematics. All contributions undergo a peer-review process to meet the highest standard of scientific literature.

Volumes in the subseries will include a collection of revised written versions of the communications or short research announcements or summaries, grouped by events or by topics. Contributing authors to the extended abstracts volumes remain free to use their own material as in these publications for other purposes (for example a revised and enlarged paper) without prior consent from the publisher, provided it is not identical in form and content with the original publication and provided the original source is appropriately credited.

Marianna Chatzakou · Michael Ruzhansky ·
Karel Van Bockstal
Editors

Direct and Inverse Problems with Applications

Extended Abstracts of the 2024 GAP Center Summer School

Birkhäuser

Editors
Marianna Chatzakou
Department of Mathematics: Analysis, Logic and Discrete Mathematics
Ghent University
Ghent, Belgium

Michael Ruzhansky
Department of Mathematics: Analysis, Logic and Discrete Mathematics
Ghent University
Ghent, Belgium

Karel Van Bockstal
Department of Mathematics: Analysis, Logic and Discrete Mathematics
Ghent University
Ghent, Belgium

ISSN 2297-0215 ISSN 2297-024X (electronic)
Trends in Mathematics
ISSN 2948-1724 ISSN 2948-1732 (electronic)
Research Perspectives Ghent Analysis and PDE Center
ISBN 978-3-031-98644-4 ISBN 978-3-031-98645-1 (eBook)
https://doi.org/10.1007/978-3-031-98645-1

This book is published under the imprint Birkhäuser, www.birkhauser-science.com by the registered company Springer Nature Switzerland AG
The registered company address is: Gewerbestrasse 11, 6330 Cham, Switzerland

Preface

This book is a volume in the Birkhäuser series *Research Perspectives Ghent Analysis and PDE Center*. This is a book series devoted to the publication of extended abstracts of seminars, conferences, workshops, and other scientific events related to the Ghent Analysis and PDE Center (GAPC). Volumes in this subseries include a collection of revised written versions of the communications or short research announcements or summaries, grouped by events or by topics.

GAPC is a research group in mathematics that focuses on different areas of analysis, as well as the study of partial differential equations (PDEs) and their applications. The topics of interest include functional analysis, Fourier analysis, noncommutative analysis, geometric analysis, partial differential equations of different types, harmonic analysis, functional inequalities, pseudo-differential operators, fractional derivatives, special functions, microlocal analysis, inverse problems, and imaging. The target audience of this book is any researcher working in the above fields. More details about the research activities of the center can be found on the group website: https://analysis-pde.org.

In addition to the research activities, the Ghent Analysis and PDE Center is also involved in training the next generation of mathematicians. GAPC organizes seminars, workshops, and conferences to share their research findings and collaborate with other researchers in the field. The center brings together researchers from different disciplines to collaborate on projects related to analysis and partial differential equations, and provides a platform for sharing knowledge and expertise.

Since 2022, GAPC has established a tradition of organizing a seasonal school every year:

- 2022: Summer School "Singularities in Science and Engineering":
  https://analysis-pde.org/summer-school-singularities-in-science-and-engineering/
- 2023: Summer School "Modern Problems in PDEs and Applications":
  https://analysis-pde.org/summer-school-modern-problems-in-pdes-and-applications/

- 2024: Summer School "Direct and Inverse Problems with Applications, and Related Topics: A Comprehensive Summer School in Mathematical Analysis": https://analysis-pde.org/summer-school-direct-and-inverse-problems-with-applications-and-related-topics/.

These schools have been very successful. A volume (containing 18 contributions of lecturers and participants about the topics they presented) in the book series *Research Perspectives Ghent Analysis and PDE Center* was devoted to the 2023 summer school:

> Chatzakou M., Restrepo J., Ruzhansky M., Torebek B., Van Bockstal K. (Eds.) Modern Problems in PDEs and Applications. Extended Abstracts of the 2023 GAP Center Summer School, Research Perspectives Ghent Analysis and PDE Center, Trends in Mathematics, Birkhäuser, 2024. viii + 190 pp.

The principal aim of this volume is to continue this initiative for the 2024 summer school. We have collected 23 contributions. In the first part of this volume, we gather contributions from the speakers of the summer school "Direct and Inverse Problems with Applications, and Related Topics" organized by our center in Ghent from August 19 to August 24, 2024. Some of the lectures that were part of the summer school were given later on, during autumn 2024. The first part contains these lectures as well. In parallel to this event, we organized two workshops and the contributions of speakers in these workshops are collected in the second part of this volume.

The aims of the school and related workshops were to explore the comprehensive mathematical tools for solving direct and inverse problems across various fields, fostering new methods and techniques through lectures, exercises, and participant interactions. Another aim of the summer school was to progress and advance in the problems considered, providing younger researchers with ideas and tools for further independent research. The topics discussed are very important since theories and methods in different areas can be developed independently, aiming to provide strong support to problems related to partial differential equations.

The summer school was organised in a friendly atmosphere and synergy, and it was an excellent opportunity to promote and encourage the development of the subject in the community and a good chance for researchers of different backgrounds and generations to interact enthusiastically and productively.

The summer school was supported by a grant from the Doctoral Schools of Ghent University. It was also partially supported by the FWO Odysseus and UGent Methusalem projects "Analysis and Partial Differential Equations."

We are grateful to all the authors who have contributed to this volume.

Ghent, Belgium
February 2025

Marianna Chatzakou
Michael Ruzhansky
Karel Van Bockstal

Contents

Summer School Courses

Fourier Algebras and Homomorphisms

Michail Anoussis

Abstract We present basic properties of the Fourier and the Fourier–Stieltjes algebras of locally compact groups and discuss the results of Cohen, Ilie-Spronk and Daws on homomorphisms of Fourier algebras. We provide necessary and sufficient conditions for the existence of idempotents of arbitrarily large norms in the Fourier-Stieltjes algebra $B(G)$ of a locally compact group G. We prove that $B(G)$ contains idempotents of arbitrarily large norm if and only if there are homomorphisms of arbitrarily large norm from the Fourier algebra $A(H)$ into $B(G)$ for any locally compact group H. The results are from joint work with Eleftherakis and Katavolos (Stud. Math. 267:109–120, 2022, [1]).

Keywords Fourier algebra · Group homomorphism · Idempotent

MSC2020 43A22

1 The Wiener Algebra

We consider the circle group $\mathbb{T} = \{z \in \mathbb{C} : |z| = 1\}$ with the parametrization $\mathbb{T} = \{e^{it} : t \in [0, 2\pi)\}$ and write $f(t)$ instead of $f(e^{it})$. The Fourier transform of $f \in L^1(\mathbb{T})$ is defined by

$$\widehat{f}(n) = \frac{1}{2\pi} \int f(t)e^{-int}dt$$

where dt is the Lebesgue measure on $[0, 2\pi)$. The Fourier series of $f \in L^1(\mathbb{T})$ is the following series:

$$\sum_{n\in\mathbb{Z}} \widehat{f}(n)e^{int}.$$

M. Anoussis (✉)
Department of Mathematics, University of the Aegean, Samos, Greece
e-mail: mano@aegean.gr

M. Chatzakou et al. (eds.), *Direct and Inverse Problems with Applications*,
Trends in Mathematics 12, https://doi.org/10.1007/978-3-031-98645-1_1

The Wiener algebra $A(\mathbb{T})$ is the space of functions $f : \mathbb{T} \to \mathbb{C}$ whose Fourier series $\sum_{n\in\mathbb{Z}} \widehat{f}(n)e^{int}$ converges absolutely. Hence, $A(\mathbb{T})$ with the norm

$$\|f\| = \sum_{n\in\mathbb{Z}} |\widehat{f}(n)| = \|\widehat{f}\|_{\ell^1(\mathbb{Z})}$$

is a Banach space isometric to $\ell^1(\mathbb{Z})$. Moreover, $A(\mathbb{T})$ is an algebra under pointwise multiplication.

Lemma 1.1 *If $f, g \in A(\mathbb{T})$, then $fg \in A(\mathbb{T})$ and $\|fg\| \leq \|f\|\|g\|$.*

Note that $A(\mathbb{T})$ with pointwise multiplication is isomorphic to the algebra $\ell^1(\mathbb{Z})$ considered with the following multiplication:

$$(a_n)_{n\in\mathbb{Z}} * (b_n)_{n\in\mathbb{Z}} = (c_n)_{n\in\mathbb{Z}}$$

where $c_n = \sum_k a_k b_{n-k}$. We have the following.

Theorem 1.1 *Let ω be a nonzero multiplicative linear functional (i.e. a character) of $A(\mathbb{T})$. There exists $z_0 \in \mathbb{T}$ such that*

$$\omega(f) = f(z_0),$$

an so the spectrum of $A(\mathbb{T})$ is canonically identified with $\mathbb{T}$.

Using Theorem 1.1. We can prove the following result of N. Wiener.

Theorem 1.2 (Wiener) *Assume that $f \in A(\mathbb{T})$ and that $f(z) \neq 0$ for all $z \in \mathbb{T}$. Then $1/f \in A(\mathbb{T})$.*

Let G be a locally compact topological group. Then G has a left-invariant measure, unique up to a scalar. This measure is called the Haar measure and is denoted by μ. We consider the Hilbert space $L^2(G)$ with inner product

$$\langle f, g\rangle = \int_G f(x)\overline{g(x)}d\mu(x).$$

The representation λ of G on $L^2(G)$ defined by

$$(\lambda(y)f)(x) = f(y^{-1}x)$$

for $f \in L^2(G)$ is called the left regular representation of the group G. Let $f \in A(\mathbb{T})$ and $g(n) \in \mathbb{C}$ defined by: $\widehat{g}(n)^2 = \widehat{f}(n)$ for all $n \in \mathbb{Z}$. Let $g(t)$ be the limit of the series $\sum_{n\in\mathbb{Z}} \widehat{g}(n)e^{int}$ in $L^2(\mathbb{T})$. We have that

$$\widehat{(g * g)}(n) = \widehat{g}(n)^2 = \widehat{f}(n)$$

and hence

$$f(t) = (g * g)(t) = \frac{1}{2\pi} \int g(t-s)g(s)ds = \langle \lambda(t)h, \bar{g} \rangle,$$

where $h(s) = g(-s)$. A function $u : G \to \mathbb{C}$ of the form

$$u(x) = \langle \pi(x)\xi, \eta \rangle$$

where π is a representation of G on a Hilbert space H and $\xi, \eta \in H$ is called a coefficient of the representation π. It follows that if $f \in A(\mathbb{T})$, then f is a coefficient of the left regular representation of the group $\mathbb{T}$.

2 Fourier and Fourier-Stieltjes Algebras

We present basic properties of the Fourier and the Fourier–Stieltjes algebras of a locally compact group. The Fourier-Stieltjes algebra $B(G)$ and the Fourier algebra $A(G)$ of a locally compact group G were introduced by Eymard [5]. The Fourier-Stieltjes algebra $B(G)$ of G is the space of all functions $u : G \to \mathbb{C}$ of the form

$$u(x) = \langle \pi(x)\xi, \eta \rangle$$

where π is a unitary representation of G on a Hilbert space H_π and ξ, η are vectors in H_π (i.e. $B(G)$ is the space of coefficients of unitary representations of the group G). It is an algebra under pointwise multiplication and is a Banach algebra with norm

$$\|u\| = \inf\{\|\xi\|.\|\eta\|\}$$

where the infimum is taken over all representations π of the group G and all ξ, η in the space of the representation H_π of π such that

$$u(x) = \langle \pi(x)\xi, \eta \rangle.$$

The Fourier algebra $A(G)$ of G is the subspace of the Fourier-Stieltjes algebra $B(G)$ of G consisting of all functions $u : G \to \mathbb{C}$ of the form

$$u(x) = \langle \lambda(x)\xi, \eta \rangle$$

where λ is the left regular representation of G and ξ, η are in $L^2(G)$. The following properties of the Fourier algebra are proved in [5]:

- The Fourier algebra $A(G)$ is a closed ideal of the Fourier-Stieltjes algebra $B(G)$. Moreover, $B(G) \cap C_c(G)$ is densely contained in $A(G)$, where $C_c(G)$ is the space of compactly supported continuous functions $u : G \to \mathbb{C}$.
- If ρ is a character of $A(G)$, then ρ is of the form $u \mapsto u(s)$ for some $s \in G$. The spectrum of the Banach algebra $A(G)$ is homeomorphic to G.
- For any closed set $K \subseteq G$ and every point $x_0 \in G$, $x_0 \notin K$ there exists $f \in A(G)$ such that $f(x) = 0$ for all $x \in K$ and $f(x_0) = 1$ (regularity).

When G is a locally compact abelian group with dual group $\widehat{G}$, the Fourier-Stieltjes algebra $B(G)$ of G is the algebra of Fourier-Stieltjes transforms of elements of the measure algebra $M(\widehat{G})$ of $\widehat{G}$ and the Fourier algebra $A(G)$ of G is the algebra of Fourier transforms of elements of $L^1(\widehat{G})$. In particular, the Fourier algebra of the group $\mathbb{T}$ is the Wiener algebra $A(\mathbb{T})$. The following theorem was proved by Walter [9].

Theorem 1.3 *Let G and H be locally compact groups. The following are equivalent:*

- *$B(G)$ and $B(H)$ are isometrically isomorphic as Banach algebras,*
- *$A(G)$ and $A(H)$ are isometrically isomorphic as Banach algebras,*
- *G and H are isomorphic as topological groups.*

3 Homomorphisms

Cohen characterized the homomorphisms from $A(H)$ into $B(G)$ in terms of piecewise affine maps when H, G are locally compact abelian groups [3]. To obtain his result he proved a characterization of idempotents in $B(G)$ [2]. Host extended the characterization of idempotents in $B(G)$ to general locally compact groups [7]. Daws [4] and Ilie and Spronk [8] characterized completely bounded homomorphisms from $A(H)$ into $B(G)$ for locally compact groups H, G with H amenable, in terms of piecewise affine maps. Let $G,\ H$ be locally compact groups and

$$\rho : A(H) \to B(G)$$

be a bounded homomorphism. Given $s \in G$, the map $\rho_s : A(H) \to \mathbb{C}$, defined by

$$\rho_s(u) = \rho(u)(s)$$

is a multiplicative linear functional on $A(H)$ (possibly 0). Let $Y = \{s \in G : \rho_s \neq 0\}$ and $s \in Y$. Then ρ_s is a character of $A(H)$ and so there is an element $\alpha(s) \in H$ such that

$$\rho(u)(s) = u(\alpha(s))$$

for all $u \in A(H)$. We obtain a map $\alpha : Y \to H$ such that

$$\rho(u) = \chi_Y(u \circ \alpha), \quad u \in A(H)$$

where χ_Y is the characteristic function of the set Y. Let G, H be locally compact groups, K be a subgroup of G, $t_0 \in G$ and $C = t_0^{-1}K$. A map $\alpha : C \to H$ is called affine if there exists a continuous homomorphism $\theta : K \to H$ and $s_0 \in H$ such that

$$\alpha(t) = s_0\theta(t_0 t),$$

for all $t \in C$. The open coset ring $\Omega_0(G)$ of a locally compact group G is the ring generated by the cosets of the open subgroups of the group G.

Theorem 1.4 ([2, 7]) *If G is a locally compact group and $F \subseteq G$, the idempotent χ_F is in the Fourier-Stieltjes algebra $B(G)$ of G if and only if $F \in \Omega_0(G)$.*

Let G, H be locally compact groups and $Y \subseteq G$. A map $\alpha : Y \to H$ is called piecewise affine if

- $Y \subseteq G$ is a disjoint union $Y = \bigcup_{i=1}^m Y_i$, where each Y_i belongs to the open coset ring $\Omega_0(G)$;
- each restriction $\alpha|_{Y_i}$ extends to an affine map $\alpha_i : C_i \to H$ defined on an open coset $C_i \supseteq Y_i$.

Let $\alpha : Y \to H$ be a piecewise affine map where $Y \subseteq G$ as before. Define $\rho : A(H) \to B(G)$ by

$$\rho(u)(t) = \begin{cases} (u \circ \alpha)(t), & t \in Y \\ 0, & t \in G \backslash Y \end{cases}. \tag{1}$$

Theorem 1.5 ([2]) *Let G, H be locally compact abelian groups and ρ as in* (1). *Then ρ is a bounded homomorphism. Every bounded homomorphism $A(H) \to B(G)$ is of this form.*

Theorem 1.6 ([4, 8]) *Let G, H be locally compact groups and ρ as in* (1). *The map ρ is a completely bounded homomorphism. If H is amenable, every completely bounded homomorphism $A(H) \to B(G)$ is of this form.*

This result generalizes the result of Cohen for the abelian case, since it is well known that abelian groups are amenable and it follows from [6] that if H is abelian any bounded linear map from $A(H)$ to another operator space is completely bounded. Moreover, Ilie and Spronk show in [8] that if H is a locally compact group which contains a discrete copy of the free group with two generators $\mathbb{F}_2$ (and hence non amenable), then there exists a completely bounded homomorphism $A(H) \to A(\mathbb{F}_2)$ which is not of the above form (1).

4 Norms of Idempotents and Homomorphisms

Let $F = \{-k, \dots, k\} \subseteq \mathbb{Z}$. For $u \in A(\mathbb{Z})$, let

$$u_F(j) := \begin{cases} u(0) & \text{if } j \in F \\ 0 & \text{if } j \notin F \end{cases}$$

Then the map

$$\rho_F : A(\mathbb{Z}) \to A(\mathbb{Z}) : u \mapsto u_F$$

is of the form $u \mapsto \chi_F(u \circ \alpha)$ where $\alpha : F \to \mathbb{Z}$ is the piecewise affine map defined by $\alpha(j) = 0$ for each $j \in F$. Hence ρ_F is a well-defined bounded homomorphism. Consider the function $u_0 : \mathbb{Z} \to \mathbb{Z}$ given by $u_0(i) = \delta_{i,0}$. Then $u_0 \in A(\mathbb{Z})$ and $\|u_0\|_{A(\mathbb{Z})} = 1$. Since $\rho_F(u_0) = \chi_F$, its Fourier transform is $\widehat{\chi_F}(t) = \sum_{n \in F} e^{-int}$ and so

$$\begin{aligned} \|\rho_F\| \geq \|\rho_F(u_0)\|_{A(\mathbb{Z})} &= \|\chi_F\|_{A(\mathbb{Z})} = \|\widehat{\chi_F}\|_{L^1(\mathbb{T})} \\ &= \frac{1}{2\pi} \int_0^{2\pi} |D_k(t)| dt \end{aligned}$$

where D_k is the Dirichlet kernel, and it is known that the L^1 norm of D_k grows like $\log k$. In the above example, we see that the existence of idempotents of large norm in $A(\mathbb{Z})$ implies the existence of homomorphisms $A(\mathbb{Z}) \to A(\mathbb{Z})$ of large norm. In [1], we study the following questions:

1. for which locally compact groups G do there exist idempotents of arbitrarily large norm in the Fourier-Stieltjes algebra $B(G)$?
2. how is the existence of idempotents of 'large' norm related to the existence of homomorphisms of arbitrarily large norm between Fourier algebras?

To prove our result, we consider the following steps.

1st step

We show that if G is an infinite totally disconnected group, then $B(G)$ contains idempotents of arbitrarily large norm.

Theorem 1.7 ([1]) *Let G be an infinite totally disconnected group. Then*

$$\sup\{\|\chi_F\|_{B(G)} : F \subseteq G,\ \chi_F \in B(G)\} = +\infty.$$

2nd step

For any locally compact group G the group G/G_e is totally disconnected (where G_e is the connected component of the identity e). Moreover, it follows from [5, 2.26,

Corollary 3] that there is an isometric embedding of $B(G/G_e)$ into $B(G)$. Using these results and Theorem 1.7, we obtain the following:

Theorem 1.8 ([1]) *Let G be a locally compact group and G_e be the connected component of $e \in G$. The following are equivalent*

1. *The quotient G/G_e is infinite.*
2. $\sup\{\|\chi_F\|_{B(G)} : \chi_F \in B(G)\} = +\infty$.

3rd step

We construct homomorphisms of large norm using idempotents of large norm.

Proposition 1.1 ([1]) *Let G, H be locally compact groups and $F \in \Omega_0(G)$. For $u \in A(H)$ we define*

$$\rho_F(u)(t) = \begin{cases} u(e), & t \in F \\ 0, & t \notin F \end{cases}$$

Then the map ρ_F is a bounded homomorphism $A(H) \to B(G)$ and

$$\|\rho_F\| = \|\chi_F\|_{B(G)}.$$

Theorem 1.9 ([1]) *Let G be a locally compact group and G_e be the connected component of $e \in G$. The following are equivalent:*

1. *The group G/G_e is infinite.*
2. $\sup\{\|\chi_F\|_{B(G)} : F \in \Omega_0(G)\} = +\infty$.
3. *For every locally compact group H,*

$$\sup\{\|\rho\|\} = +\infty.$$

where the sup is taken over all bounded homomorphisms $\rho : A(H) \to B(G)$.

References

1. Anoussis, M., Eleftherakis, G.K., Katavolos, A.: Idempotents of large norm and homomorphisms of Fourier algebras. Stud. Math. **267**, 109–120 (2022)
2. Cohen, P.J.: On a conjecture of Littlewood and idempotent measures. Amer. J. Math. **82**, 191–212 (1960)
3. Cohen, P.J.: On homomorphisms of group algebras. Amer. J. Math. **82**, 213–226 (1960)
4. Daws, M.: Completely bounded homomorphisms of the Fourier algebra revisited. J. Group Theory **25**, 579–600 (2022)
5. Eymard, P.: L'algèbre de Fourier d'un groupe localement compact. Bull. Soc. Math. France **92**, 181–236 (1964)
6. Forrest, B., Wood, P.: Cohomology and the operator space structure of the Fourier algebra and its second dual. Indiana Univ. Math. J. **50**, 1217–1240 (2001)

7. Host, B.: Le théorème des idempotents dans $B(G)$. Bull. Soc. Math. France **114**, 215–223 (1986)
8. Ilie, M., Spronk, N.: Completely bounded homomorphisms of the Fourier algebras. J. Funct. Anal. **225**, 480–499 (2005)
9. Walter, M.E.: W^*-algebras and nonabelian harmonic analysis. J. Funct. Anal. **11**, 17–38 (1972)

Dispersion Phenomena and Applications to Evolution Equations

Hajer Bahouri

Abstract The aim of these lectures is to give an overview of dispersion phenomena on the Euclidean space $\mathbb{R}^d$, as well as on some non-commutative Lie groups. Then, these properties will be applied to investigate some nonlinear evolution equations.

Keywords Dispersive inequalities · Strichartz estimates · Fourier restriction · Kato smoothing effect

MSC2020 43A30 · 43A80

1 Introduction

Dispersion phenomena for evolution equations (such as the Schrödinger and wave equations) express that waves u with different frequencies move at different velocities, and correspond commonly to pointwise inequalities in time decay, under the form ($t \neq 0$)

$$\|u(t,\cdot)\|_{L^\infty} \lesssim \frac{\|u_0\|_{L^1}}{|t|^r},$$

where (in general) the rate of decay $r > 0$ depends on the equation, the dimension and the setting. In the $\mathbb{R}^d$ framework, dispersive inequalities have a long history, beginning with Brenner [7], Pecher [17], Segal [19] and Strichartz [22]. In many cases, the main step in the proof of this decay in time relies on the application of a stationary phase theorem on an (approximate) representation of the solution.

Combined with an abstract functional analysis argument known as the TT^*-argument (developed by Ginibre and Velo [11] and refined by Keel and Tao [14]), dispersion phenomena yield a range of estimates involving space-time Lebesgue norms. Those inequalities, called Strichartz estimates, have proved to be powerful in

H. Bahouri (✉)
Laboratoire Jacques-Louis Lions (LJLL), CNRS and Sorbonne Université, Paris, France
e-mail: hajer.bahouri@sorbonne-universite.fr

M. Chatzakou et al. (eds.), *Direct and Inverse Problems with Applications*,
Trends in Mathematics 12, https://doi.org/10.1007/978-3-031-98645-1_2

the study of semilinear and quasilinear problems which appear in numerous physical applications.

When some loss of dispersion occurs, such as for instance in the case of compact Riemannian manifolds, or in the case of the Schrödinger operator on the Heisenberg group (where it was shown by Bahouri-Gérard et al. [5] that there is no dispersion at all), obtaining Strichartz estimates is a very difficult issue. However, other approaches, such as the Fourier transform restriction method [23], the Kato smoothing effect [13] or the microlocal analysis can be successfully used to obtain Strichartz estimates (with a possible loss of derivatives).

The structure of this text is as follows. In Sect. 2, we record some basic dispersive inequalities concerning the Schrödinger and wave equations on $\mathbb{R}^d$, while in Sect. 3 we briefly introduce other approaches that can be used when dispersion phenomena fail.

2 The Euclidean Case

2.1 Dispersive Estimates

For the Schrödinger equation on $\mathbb{R}^d$

$$(S) \qquad \begin{cases} i\partial_t u - \Delta u = 0 \\ u_{|t=0} = u_0, \end{cases}$$

the dispersive estimate writes ($t \neq 0$)

$$\|u(t,.)\|_{L^\infty} \leq \frac{\|u_0\|_{L^1}}{(4\pi|t|)^{\frac{d}{2}}}. \tag{2.1}$$

Indeed, taking the partial Fourier transform $\mathcal{F}$ of (S) with respect to the variable x, we obtain

$$i\partial_t \mathcal{F}u(\xi) + |\xi|^2 \mathcal{F}u(\xi) = 0.$$

Then integrating in time the resulting ODE, we get

$$\mathcal{F}u(t,\xi) = e^{it|\xi|^2}\mathcal{F}(u_0)(\xi).$$

Applying the inverse Fourier formula, we obtain (oscillating integral)

$$u(t,x) = (2\pi)^{-d} \int_{\mathbb{R}^d} e^{i(x\cdot\xi + t|\xi|^2)} \mathcal{F}(u_0)(\xi) d\xi = \Big(\frac{e^{i\frac{|\cdot|^2}{4t}}}{(4\pi i t)^{\frac{d}{2}}} \star u_0 \Big)(x),$$

which, thanks to Young inequality, completes the proof of (2.1).

Remark 2.1 Note that using the stationary phase theorem, one can show that this estimate is sharp, that is to say that the rate of decay $\frac{d}{2}$ cannot be improved. Moreover, since $\|u(t,\cdot)\|_{L^2} = \|u_0\|_{L^2}$, a complex interpolation argument (Riesz-Thorin's theorem) leads to the following family of dispersive inequalities, for $2 \le p \le \infty$,

$$\|u(t,\cdot)\|_{L^p} \lesssim \frac{1}{|t|^{\frac{d}{2}-\frac{d}{p}}}\|u_0\|_{L^{p'}},$$

where p' denotes the conjugate exponent of p.

Concerning the wave equation on $\mathbb{R}^d$

$$(W)\begin{cases} \partial_t^2 u - \Delta u = 0 \\ u_{|t=0} = u_0 \\ (\partial_t u)_{|t=0} = u_1, \end{cases}$$

for Cauchy data frequency localized in a unit ring C, we have the following dispersive estimate[1]

$$\|u(t,\cdot)\|_{L^\infty} \lesssim |t|^{-\frac{d-1}{2}}\big(\|u_0\|_{L^1} + \|u_1\|_{L^1}\big), \tag{2.2}$$

which is also sharp by the stationary phase theorem. Indeed, applying Fourier analysis, one can write $u = u_+ + u_-$ with

$$\widehat{u}_\pm(t,\xi) = \mathrm{e}^{\pm it|\xi|}\varphi(\xi)\widehat{\gamma}_\pm(\xi)$$

where $\varphi \in \mathcal{D}(\mathbb{R}^d\backslash\{0\})$ with $\varphi \equiv 1$ near the unit ring C. Then by inversion formula, we find that $u_\pm(t,\cdot) = K_\pm(t,\cdot) \star \gamma_\pm$ with

$$K_\pm(t,x) = (2\pi)^{-d}\int \mathrm{e}^{i(x\cdot\xi \pm t|\xi|)}\varphi(\xi)d\xi.$$

By virtue of Young inequality, we get

$$\|u_\pm(t,\cdot)\|_{L^\infty} \le \|K_\pm(t,\cdot)\|_{L^\infty}\|\gamma_\pm\|_{L^1}.$$

Since $\gamma_\pm = \frac{1}{2}\big(u_0 \mp ih \star u_1\big)$, where $\widehat{h} = \frac{\varphi}{|\cdot|}$, one can easily check that

$$\|\gamma_\pm\|_{L^1} \lesssim \|u_0\|_{L^1} + \|u_1\|_{L^1},$$

and thus the problem reduces to the estimate of the L^∞-norm of $K_\pm(t,\cdot)$. For that purpose, we apply the stationary phase theorem (integration by parts with respect to a suitable vector field) to

[1] Which is weaker than the estimate (2.1) corresponding to the Schrödinger equation on $\mathbb{R}^d$.

$$K_{\pm}(t, X) = (2\pi)^{-d} \int e^{it\Phi_{\pm}(X,\xi)} \varphi(\xi) d\xi,$$

where we denote $\Phi_{\pm}(X, \xi) = X \cdot \xi \pm |\xi|$. The study of oscillatory integrals is in connection with the critical point of the phase, here $\Phi_{\pm}$. So, we split $K_{\pm}$ on two parts as follows:

$$(2\pi)^d K_{\pm}(t, X) = \int_{\mathbb{R}^d \backslash C_X} e^{it\Phi_{\pm}(X,\xi)} \varphi(\xi) d\xi + \int_{C_X} e^{it\Phi_{\pm}(X,\xi)} \varphi(\xi) d\xi$$

with $C_X = \left\{\xi \in C / \left|X \pm \frac{\xi}{|\xi|}\right| \leq \frac{1}{2}\right\}$.

First on $\mathbb{R}^d \backslash C_X$, we have $|\nabla \Phi_{\pm}| \geq \frac{1}{2}$ (the non stationary phase case). Then, integrating by parts with respect to the vector field $\mathcal{L}_{\pm}$ defined by

$$\mathcal{L}_{\pm} a = -i \frac{\nabla \Phi_{\pm}}{|\nabla \Phi_{\pm}|^2} \cdot \nabla a$$

and which satisfies $\mathcal{L}_{\pm} e^{it\Phi_{\pm}} = t e^{it\Phi_{\pm}}$, one can easily show that, for any integer N,

$$\Big| \int_{\mathbb{R}^d \backslash C_X} e^{it\Phi_{\pm}(X,\xi)} \varphi(\xi) d\xi \Big| \leq \frac{C_N}{|t|^N}.$$

Now on C_X, we have $|\nabla \Phi_{\pm}| \leq \frac{1}{2}$ (the stationary phase case). Performing integrations by parts with respect to the vector field $\tilde{\mathcal{L}}_{\pm}$ (which also satisfies $\tilde{\mathcal{L}}_{\pm} e^{it\Phi_{\pm}} = e^{it\Phi_{\pm}}$)

$$\tilde{\mathcal{L}}_{\pm} = \frac{1}{1 + t|\nabla \Phi_{\pm}|^2} (Id - i\nabla \Phi_{\pm} \cdot \nabla),$$

we find that

$$\Big| \int_{C_X} e^{it\Phi_{\pm}(X,\xi)} \varphi(\xi) d\xi \Big| \lesssim \int_{C_X} \frac{d\xi}{\left(1 + t\left|X \pm \frac{\xi}{|\xi|}\right|^2\right)^d}.$$

If C_X is not empty, then $X \neq 0$, and one can write the following orthogonal decomposition for any $\xi \in C_X$:

$$\xi = \xi_1 + \zeta' \quad \text{with} \quad \xi_1 = \Big(\xi \Big| \frac{X}{|X|}\Big) \frac{X}{|X|} \quad \textit{and} \quad \zeta' = \xi - \Big(\xi \Big| \frac{X}{|X|}\Big) \frac{X}{|X|}.$$

Since ζ' is orthogonal to the vector X, we have $\left|X \pm \frac{\xi}{|\xi|}\right| \geq \frac{|\zeta'|}{|\xi|}$, which implies that

$$\Big| \int_{C_X} e^{it\Phi_{\pm}(X,\xi)} \varphi(\xi) d\xi \Big| \lesssim \int_{C} \frac{1}{\left(1 + t|\zeta'|^2\right)^d} d\zeta' d\xi_1.$$

Considering the change of variables $\tilde{\zeta} = \sqrt{t}\,\zeta'$, one can easily check that

$$\Big|\int_{C_X} e^{it\Phi_\pm(X,\xi)}\varphi(\xi)d\xi\Big| \lesssim \frac{1}{|t|^{\frac{d-1}{2}}}.$$

This completes the proof of the dispersive estimate (2.2).

Remark 2.2 Note that Strichartz estimates in regular or rough variable coefficients frameworks as well as in the quasilinear framework (for short time) have been obtained in a number of works. There are also various studies in the setting of compact Riemannian manifolds and bounded domains. For all these works requiring more elaborate techniques, such as Littlewood-Paley theory,[2] one can see for instance [2, 8, 12, 15, 20, 21] and the references therein.

2.2 *Strichartz Estimates*

A functional analysis argument known as the TT^*-argument initiated by Ginibre and Velo [11] and refined by Keel and Tao [14] enables to deduce from the family of dispersive estimates a bound for the space-time norm of the solutions (both in the homogeneous and inhomogeneous cases) by means of the norm of the data. These estimates known as Strichartz estimates (which depend on the equation and the dimension) constitute a central tool in the study of nonlinear equations, whether in semilinear frameworks or in quasilinear settings. This abstract argument can be stated as follows (see, for instance, Chap. 8 in [1] for the proof).

Theorem 2.1 *Let $U(t)$ a bounded family of operators on $L^2(\mathbb{R}^d)$ such that, for some $\sigma > 0$*

$$\big\|U(t)U^\star(t')f\big\|_{L^\infty(\mathbb{R}^d)} \lesssim \frac{1}{|t-t'|^\sigma}\|f\|_{L^1(\mathbb{R}^d)}.$$

Then if

$$\frac{2}{p}+\frac{2\sigma}{q}=\sigma, \quad (\sigma,p,q)\neq(1,2,\infty),$$

we have

$$\big\|U(t)u_0\big\|_{L^p(\mathbb{R};L^q(\mathbb{R}^d))} \lesssim \|u_0\|_{L^2(\mathbb{R}^d)},$$

$$\Big\|\int_0^t U(t)U^\star(t')f(t')\,dt'\Big\|_{L^p(\mathbb{R};L^q(\mathbb{R}^d))} \lesssim \|f\|_{L^{p_1'}(\mathbb{R};L^{q_1'}(\mathbb{R}^d))}.$$

This theorem plays a key role in the analysis of nonlinear evolution equations. As an application, we here show how to solve the cubic Schrödinger equation on $\mathbb{R}^2$:

[2] For an introduction to this topic, one can consult [6].

$$(NLS_3) \qquad \begin{cases} i\partial_t u - \Delta u = P_3(u, \overline{u}) \\ \qquad u_{|t=0} = u_0 \in L^2(\mathbb{R}^2), \end{cases}$$

where P_3 is an homogeneous polynomial of degree 3 with respect to u and $\overline{u}$ (for instance, one can take $P_3(u, \overline{u}) = |u|^2 u$). Observing that $L^3_t(L^6_x)(\mathbb{R}^2)$ is a Strichartz norm, one can easily show thanks to the fixed point theorem that (NLS_3) is locally (in time) well-posed for any Cauchy data in L^2 and even globally (in time) well-posed for small Cauchy data. More precisely, we have the following theorem.

Theorem 2.2 *Let $u_0 \in L^2(\mathbb{R}^2)$. Then there exists a positive time $T = T(\|u_0\|_{L^2})$ such that there exists a unique solution u of (NLS_3) in the functional space $C([0, T]; L^2(\mathbb{R}^2)) \cap L^3([0, T]; L^6(\mathbb{R}^2))$. Moreover, there is a positive constant $c > 0$ such that if $\|u_0\|_{L^2} \leq c$, then the solution belongs to $L^3(\mathbb{R}_+; L^6(\mathbb{R}^2)) \cap C_b(\mathbb{R}_+; L^2(\mathbb{R}^2))$.*

To prove this result, we start by observing that a solution of (NLS_3) is a fixed point of the map $F\colon u \longrightarrow F(u) = u_F + Q(u)$, with

$$(i\partial_t - \Delta)u_F = 0, \; u_{F\,|t=0} = u_0$$

and

$$(i\partial_t - \Delta)Q(u) = P_3(u, \overline{u}), \; Q(u)_{|t=0} = 0.$$

By Strichartz estimates (here $L^3(L^6)$), we have for any positive time T

$$\|F(u)\|_{L^3([0,T],L^6(\mathbb{R}^2))} \leq C(\|u_0\|_{L^2(\mathbb{R}^2)} + \|u\|^3_{L^3([0,T],L^6(\mathbb{R}^2))}),$$

and

$$\|F(u) - F(v)\|_{L^3([0,T],L^6(\mathbb{R}^2))} \leq C\|u_0\|^2_{L^2(\mathbb{R}^2)}\|u - v\|^3_{L^3([0,T],L^6(\mathbb{R}^2))},$$

which by standard fixed point theorem imply the result.

3 Other Alternatives

The setting has a big influence on dispersion effects, and there are many situations where dispersion phenomena fail, such as for instance in compact manifolds, on some domains, or on the Heisenberg group in the case of the Schrödinger equation. In this section, we briefly introduce the Fourier-restriction and Kato-smoothing approaches that could bypass the lack of dispersion.

3.1 Fourier-Restriction Approach

The founding result in this regards, known as Tomas-Stein theorem, can be stated as follows, where the index p_0 is optimal according to a basic counterexample known as Knapp's counterexample.

Theorem 2.3 *Let $d \geq 2$ and $p_0 = \frac{2(d+1)}{d+3}$. Let S be a smooth and compact hypersurface in $\mathbb{R}^d$ endowed with a smooth measure $d\sigma$. Assume that S has non-vanishing Gaussian curvature at every point. Then for all $p \in [1, p_0]$, there exists a positive constant $C = C_p$ such that for any $f \in \mathcal{S}(\mathbb{R}^d)$:*

$$\|(\mathcal{F} f)_{|S}\|_{L^2(S,d\sigma)} \leq C\|f\|_{L^p(\mathbb{R}^d)}.$$

Let us show how this theorem enables to recover Strichartz estimates for instance for the Schrödinger equation (S) on $\mathbb{R}^d$. We have seen that

$$u(t,x) = (2\pi)^{-d} \int_{\mathbb{R}^d} e^{i(x\cdot\xi + t|\xi|^2)} \mathcal{F}(u_0)(\xi) d\xi.$$

This formula can be interpreted as the restriction of the $\mathbb{R}^{1+d}$-Fourier transform on the paraboloid S in the space of frequencies of $\mathbb{R}^{1+d}$, defined as

$$S = \Big\{(\alpha, \xi) \in \mathbb{R}^{1+d} \mid \alpha = |\xi|^2\Big\}.$$

More precisely, setting $y = (t, x)$ and $z = (\alpha, \xi)$, we infer that

$$u(t,x) = (2\pi)^{-d} \int_{S} e^{iy\cdot z} g(z) d\sigma(z),$$

where $g(|\xi|^2, \xi) = \mathcal{F}(u_0)(\xi)$, which easily implies by the dual form of Tomas-Stein theorem that

$$\|u\|_{L^{2+\frac{4}{d}}(\mathbb{R}, L^{2+\frac{4}{d}}(\mathbb{R}^d))} \lesssim \|u_0\|_{L^2(\mathbb{R}^d)}.$$

Tomas-Stein theorem admits the following counterpart on the Heisenberg group $\mathbb{H}^d$ (see for instance [9, 10], for an introduction to the Heisenberg group) due to Müller [16].

Theorem 2.4 *For $1 \leq p \leq 2$*

$$\|\mathcal{F}_{\mathbb{H}}(f)\|_{|\widehat{S}}\|_{L^2(\widehat{S})} \leq C\|f\|_{L^p_Y L^1_s},$$

where $w = (Y, s) \in \mathbb{R}^{2d+1}$ denotes a generic element of $\mathbb{H}^d$, and $\widehat{S}$ is in some sense the unit sphere in the frequency space of $\mathbb{H}^d$, see [4] for further details.

Remark 2.3 As we can observe, there is no gain with respect to the variable s (known as the vertical variable), and this is optimal, but the result is better than the Euclidean setting with respect to variable Y, called horizontal variable. Based on Müller's restriction theorem, weak Strichartz estimates have been obtained in [3] for the Schrödinger equation on the Heisenberg group equation which is a totally non dispersive equation.

3.2 Kato-Smoothing Approach

Kato-smoothing property was established by Kato [13] for the Korteweg-de Vries equation, based on Fourier restriction type results. This property was generalized to other equations and systems by several authors. In the case of the Schrödinger equation (S), even though $\|u(t,\cdot)\|_{L^2} = \|u_0\|_{L^2}$ one can locally gain one half derivative in the following sense (one can consult the monograph of Robbiano [18] for a proof).

Theorem 2.5 *There exists a positive constant C such that, for all $u_0 \in L^2(\mathbb{R}^d)$, u the solution of the Schrödinger equation (S) satisfies*

$$\|\langle x\rangle^{-1}\langle D_x\rangle^{\frac{1}{2}}u\|_{L^2(\mathbb{R}\times\mathbb{R}^d)} \leq C\|u_0\|_{L^2(\mathbb{R}^d)}.$$

References

1. Bahouri, H., Chemin, J.-Y., Danchin, R.: Fourier analysis and nonlinear partial differential equations. Grundlehren der mathematischen Wissenshaften **343**. Springer (2011)
2. Bahouri, H., Chemin, J.-Y.: Équations d'ondes quasilinéaires et inégalités de Strichartz. Am. J. Math. **121**, 1337–1377 (1999)
3. Bahouri, H., Barilari, D., Gallagher, I.: Strichartz estimates and Fourier restriction theorems on the Heisenberg group. J. Fourier Anal. Appl. **27**, 41 (2017)
4. Bahouri, H., Chemin, J.-Y., Danchin, R.: A frequency space for the Heisenberg group. Annales de l'Institut de Fourier **69**, 365–407 (2019)
5. Bahouri, H., Gérard, P., Xu, C.-J.: Espaces de Besov et estimations de Strichartz généralisées sur le groupe de Heisenberg. Journal d'Analyse Mathématique **82**, 93–118 (2000)
6. Bahouri, H.: The Littlewood-Paley theory: a common thread of many works in nonlinear analysis. Eur. Math. Soc. Newslett. **112**, 15–23 (2019)
7. Brenner, P.: On $L^p - L^{p'}$ estimates for the wave-equation. Mathematische Zeitschrift **145**, 251–254 (1975)
8. Burq, N., Gérard, P., Tzvetkov, N.: Strichartz inequalities and the nonlinear Schrödinger equation on compact manifolds. Am. J. Math. **126**, 569–605 (2004)
9. Faraut, J., Harzallah, K.: Deux cours d'analyse harmonique. École d'Été d'analyse harmonique de Tunis. Progress in Mathematics, Birkhäuser (1984)
10. Fischer, V., Ruzhansky, M.-V.: A pseudo-differential calculus on the Heisenberg group. Comptes Rendus. Mathématique **352**, 197–204 (2014)
11. Ginibre, J., Velo, G.: Generalized Strichartz inequalities for the wave equations. J. Funct. Anal. **133**, 50–68 (1995)
12. Kapitanski, L.: Some generalization of the Strichartz-Brenner inequality. Leningrad Math. J. **1**, 693–721 (1990)

13. Kato, T.: On the Cauchy problem for the (generalized) Kortewegde Vries equation. Studies in applied mathematics. Adv. Math., Supplementary Stud. **8**, 93–128 (1983)
14. Keel, M., Tao, T.: Endpoint Strichartz estimates. Am. J. Math. **120**, 955–980 (1998)
15. Klainerman, S., Rodnianski, I., Szeftel, J.: The resolution of the bounded L^2 curvature conjecture in general relativity. Bull. Braz. Math Soc. **47**, 445-456 (2016)
16. Müller, D.: A restriction theorem for the Heisenberg group. Ann. Math. **131**, 567–587 (1990)
17. Pecher, H.: L^p—Abschätzungen und klassische Lösungen für nichtlineare Wellengleichungen. Mathematische Zeitschrift **150**, 159–183 (1976)
18. Robbiano, L.: Effets régularisants pour l'équation de Schrödinger (French) [Smoothing effects for the Schrödinger equation]. *PDE's, dispersion, scattering theory and control theory. Société Mathématique de France, Paris***30**, 1–44 (2017)
19. Segal, I.-E.: Space-time decay for solutions of wave equations. Adv. Math. **22**, 304–311 (1976)
20. Smith, H.: A parametrix construction for wave equation with $C^{1,1}$ coefficients. Annales de l'Institut Fourier **48**, 797–835 (1998)
21. Smith, H., Tataru, D.: Sharp local well-posedness results for the nonlinear wave equation. Ann. Math. **162**, 291–366 (2005)
22. Strichartz, R.: Restriction Fourier transform of quadratic surfaces and decay of solutions of the wave equations. Duke Math. J. **44**, 705–714 (1977)
23. Tomas, P.A.: A restriction theorem for the Fourier transform. Bull. Am. Math. Soc. **81**, 477–478 (1975)

Subriemannian Geometry and Analysis of Hypoelliptic PDE

Wolfram Bauer

Abstract Subriemannian (SR) geometry provides a mathematical model for motion under non-holonomic constraints. Concrete problems in mathematics, physics or applications can often be formulated in the Subriemannian context. After providing the basic definitions we recall some examples and constructions of SR geometries. Relevant questions concern horizontal connectivity as well as the classification of SR geodesics and exponential map. From an analytic point of view we recall the construction of an induced geometric differential operators called the (intrinsic) sub-Laplacian which fulfills subelliptic estimates and therefore is hypoelliptic. The corresponding SR heat equation and its fundamental solution K (SR heat kernel) establish a link between geometry and analysis and in some cases K encodes the spectrum of the operator. If available, one is interested in explicit formulas for the heat kernel. In the case of a compact manifold, eigenvalue asymptotics or the inverse spectral problem in this non-elliptic configuration can be studied.

Keywords Bracket generating distributions · Popp's measure · Intrinsic sub-Laplacian · SR heat equation

MSC2020 53C17 · 58J35

1 Definition and Examples

The present work is an extended abstract of a course given by the author at the University of Ghent, Belgium in the summer 2024. Without providing detailed proofs we survey some of the results from this lecture at the intersection of analysis and geometry. Most of the material is not original and for further details we refer to the monographs [1, 12, 20] and the seminal paper [22, 23].

W. Bauer (✉)
Institut für Analysis, Hannover, Germany
e-mail: bauer@math.uni-hannover.de

M. Chatzakou et al. (eds.), *Direct and Inverse Problems with Applications*, Trends in Mathematics 12, https://doi.org/10.1007/978-3-031-98645-1_3

Subriemannian (SR) geometry models motions under non-holonomic constraints, i.e. constraints that depend on the velocity of a curve in a manifold of configurations. We recall the precise definition.

Definition 3.1 (*Subriemannian manifold*) A *Subriemannian manifold* (shortly; SR manifold) is a triple $(M, \mathcal{H}, \langle\cdot,\cdot\rangle)$, where

1. M denotes a smooth manifold without boundary of dimensions $\dim M \geq 3$ and $\mathcal{H} \subset TM$ is a vector distribution.
2. the distribution $\mathcal{H}$ is *bracket generating* of rank $k \leq \dim M$. This means

$$\mathrm{Lie}_q(\mathcal{H}) = T_qM \quad q \in M,$$

   where the left hand side denotes the Lie hull of vector fields defined locally near $q \in M$ and taking values in $\mathcal{H}$ (we call them *horizontal*).
3. $\langle\cdot,\cdot\rangle_q$ with $q \in M$ denotes a smoothly varying family of inner products on $\mathcal{H}_q$.

Example 1 We provide some examples of SR manifolds M. More cases and some of the relevant references are found in [1, 12, 20].

1. Each Riemannian manifold (M, g) is a SR manifold with $\mathcal{H} = TM$ and $\langle\cdot,\cdot\rangle = g$ being the Riemannian metric.
2. $M := \mathbb{R}^2 \times \mathbb{S}^1$ with coordinates (x, y, ϑ) and $\mathcal{H} := \ker\omega$, where ω is the contact one-form $\omega = -\sin\vartheta\, dx + \cos\vartheta\, dy$, serves as the configuration space of a simple car-parking model. More precisely, a point $(x, y, \vartheta) \in \mathbb{R}^2 \times \mathbb{S}^1$ represents the coordinate (x, y) of the center of the car in the plane and ϑ denotes the angular between the direction of travel and the horizontal axis. An inner product on $\mathcal{H}$ may be defined by declaring a (particular) basis of horizontal vector fields spanning $\mathcal{H}$ pointwise orthonormal.
3. The three-dimensional Heisenberg group $(\mathbb{H}_3, *)$ equipped with a left-invariant SR structure is a fundamental model in SR geometry, see [20]. This is a particular case of a *left-invariant SR structure* on a Lie group G, defined in the following way: Let $V \subset \mathfrak{g}$ be a subspace of the corresponding Lie algebra $\mathfrak{g}$ with a given inner product $\langle\cdot,\cdot\rangle_V$ on V and such that V Lie generates $\mathfrak{g}$, i.e.:

$$\mathfrak{g} = \mathrm{Lie}\,(V) = \mathrm{span}\Big\{v, [w, x], \big[y, [w, x]\big], \ldots\, : \, x, y, w \in V\Big\}.$$

   Identify V via left-translation with a space of *left-invariant vector fields* on G which spans a distribution $\mathcal{H}$. Then G becomes a SR manifold $(G, \mathcal{H}, \langle\cdot,\cdot\rangle)$ with metric:

$$\langle\cdot,\cdot\rangle_g = \Big\langle (dL_g)^{-1}\cdot, (dL_g)^{-1}\cdot \Big\rangle_V.$$

   L_g denotes the left-multiplication on G by $g \in G$.
4. Example 2 and the Heisenberg group $\mathbb{H}_3$ in 3. fall into the larger class of contact manifolds: let ω be a contact one-form on M, where $\dim M = 2k + 1$, $k \in \mathbb{N}$ is

odd. Then $\mathcal{H} := \ker \omega$ defines a bracket generating distribution. After fixing an inner product $\langle \cdot, \cdot \rangle$ on $\mathcal{H}$ we obtain a SR manifold $(M, \ker \omega, \langle \cdot, \cdot \rangle)$.

5. Trivializable SR structures on (certain) Euclidean spheres $\mathbb{S}^{2k+1}$ of odd dimension (i.e. the distribution $\mathcal{H}$ is trivial as a vector bundle) have been defined and their SR spectral theory has been studied in [6, 8, 10]. Most of theses structures are not of contact type. SR structures of contact and non-contact type have been considered on *exotic spheres* as well, see [4].

On a SR manifold $(M, \mathcal{H}, \langle \cdot, \cdot \rangle)$ we consider horizontal objects, such as horizontal curves $\gamma : I \subset \mathbb{R} \to M$, i.e. $\dot{\gamma}(t) \in \mathcal{H}_{\gamma(t)}$ for a.e. $t \in I = \textit{interval}$. We may assume that the curve γ is C^1 or just absolutely continuous. On the set of horizontal curves γ we define the standard length functional

$$\ell(\gamma) := \int_I \sqrt{\langle \dot{\gamma}(t), \dot{\gamma}(t) \rangle_{\gamma(t)}} dt.$$

After a reparametrization of γ we may assume that $I = [0, 1]$. The *SR distance* between two points $a, b \in M$ is given by:

$$d_{\mathrm{cc}}(a, b) := \inf \big\{ \ell(\gamma) \ : \ \gamma \textit{ horizontal}, \ \ \gamma(0) = a, \gamma(1) = b \big\}.$$

Similarly, we can define the notation of *horizontal vector fields* X on M by requiring that $X_q \in \mathcal{H}_q$ for all $q \in M$.

It is a natural question whether on a connected SR manifold one can connect any pair of points by horizontal curves. A seminal result which independently has been obtained by Chow and Rashevsky [14, 21] (see also [1, 17]) and essentially relies on the brackets generating condition provides a positive answer.

Theorem 3.1 *Any two points on a connected SR-manifold M can be connected by piecewise smooth horizontal curves and, in particular, d_{cc} defines a metric on M. The topology on M induced by d_{cc} coincides with the manifold topology.*

The study of (locally) length minimizing horizontal curves (called SR geodesiscs) has stimulated some research in recent years and is closely linked to problems in control theory. New effects could be observed that have no equivalent in the area of Riemannian geometry and a number of problems remain unsolved to this day. As for further details we refer to [17, 19, 20] and the literature therein.

Let $M = (M, \mathcal{H}, \langle \cdot, \cdot \rangle)$ be a SR manifold. To each $q \in M$ we can assign a family of vector spaces $\mathcal{H}_q^j$ as follows:

$$\mathcal{H}_q^1 := \mathcal{H}_q, \quad \textit{and} \quad \mathcal{H}_q^{r+1} = \mathcal{H}_q^r + \big[\mathcal{H}^r, \mathcal{H}\big]_q,$$

where one defines $[\mathcal{H}^r, \mathcal{H}]_q := \mathrm{span}\{[X, Y]_q \ : \ X \in \mathcal{H}^r \ \ \textit{and} \ \ Y \in \mathcal{H}\}$. This induces a *flag* of vector spaces

$$\mathcal{H}_q = \mathcal{H}_q^1 \subset \mathcal{H}_q^2 \subset \cdots \subset \mathcal{H}_q^r \subset \mathcal{H}_q^{r+1} \subset \cdots$$

The distribution $\mathcal{H}$ being bracket generating means: for all $q \in M$ there is a minimal $\ell_q \in \mathbb{N}$ with $\mathcal{H}_q^{\ell_q} = T_q M$. Note that $\mathcal{H}_q^j$ for $j \geq 2$ in general may not be locally trivial and therefore does not define a vector bundle (cf. the Martinet distribution below).

Definition 3.2 The distribution $\mathcal{H}$ is called *regular* at $q \in M$, if $\ell_q \equiv \ell$ and the dimensions $\dim \mathcal{H}_q^r$ for all $r = 1, \ldots, \ell$ are constant in a neighbourhood of q. We call M *equiregular* if these quantities are constant globally on M.

Example 2 All SR manifolds in Example 1 are equiregular. The *Martinet distribution*

$$\mathcal{H} = \operatorname{span}\left\{X := \partial_x, +\frac{y^2}{2}\partial_z, Y := \partial_y\right\} \tag{1}$$

on $M = \mathbb{R}^3$ with coordinates (x, y, z) defines a non-equiregular SR manifold. In fact, $\mathcal{H}$ is not regular at any point on the hypersurface $\{y = 0\} \subset M$.

For each $q \in M$ we obtain a graded vector space which is called *nilpotentization* at q:

$$\mathrm{gr}(\mathcal{H})_q = \mathcal{H}_q \oplus \mathcal{H}_q^2/\mathcal{H}_q \oplus \cdots \oplus \mathcal{H}_q^{\ell_q}/\mathcal{H}_q^{\ell_q - 1}.$$

The nilpotentization has the structure of a nilpotent Lie algebra. The corresponding connected, simply connected Lie group G_q is a Carnot Lie group and carries a left-invariant SR structure itself, cf. 3. of Example 1. In fact, one observes:

(a) Lie brackets of vector fields on M induce a *Lie algebra structure* on $\mathrm{gr}(\mathcal{H})_q$ *(respecting the grading)*, see [1, 20]
(b) The Lie algebra in (a) is *nilpotent*, i.e. there is $n \in \mathbb{N}$ such that:

$$\Big[X_1[X_2 \cdots [X_n, X] \cdots]\Big] = 0, \quad \forall\ X_1, \ldots, X_n, X \in \mathrm{gr}(\mathcal{H})_q. \tag{$*$}$$

The minimal n with $(*)$ is called the *step* of the nilpotent Lie algebra.

We may think of the SR manifold G_q as a local model at a point $q \in M$ to the SR manifold. In particular, such groups play a role in the analysis of certain hypoelliptic geometric differential operators as defined in the next section (see also the monographs [20] for a precise statement).

2 The Intrinsic Sub-Laplacian

The theory of hypoelliptic operators has been developed since the 70th. In recent years and in the above framework of SR geometry it has seen some revival. In fact, PDEs induced by SR geometric structures nowadays arise in many different areas of mathematics, physics or even applied sciences including control theory, hypoelliptic diffusion, geometric measure theory, potential analysis, optimal transport or harmonic analysis. More details on the material present in the present section can be found in [1, 2, 7, 8, 20].

We recall the definition of the intrinsic sub-Laplacian Δ_{sub} on an equiregular SR manifold $(M, \mathcal{H}, \langle\cdot,\cdot\rangle)$ in [2]. Let ω be a fixed smooth measure (volume form) on M.

Definition 3.3 Given a vector field X on M and $\varphi \in C^\infty(M)$ the *horizontal gradient* and the *ω-divergence* are defined through the following equations

$$\mathcal{L}_X(\omega) = \operatorname{div}_\omega(X)\omega \quad \textit{and} \quad \Big\langle \operatorname{grad}_{\mathcal{H}}(\varphi), v \Big\rangle_q = d\varphi(v), \quad v \in \mathcal{H}_q, \quad q \in M.$$

Here $\mathcal{L}_X$ denotes the Lie derivative in direction X and $\operatorname{grad}_{\mathcal{H}}(\varphi) \in \mathcal{H}_q$ is the unique horizontal vector representing the restriction of the differential $d\varphi$ to the horizontal space $\mathcal{H}_q$.

The *sub-Laplacian* on M associated to ω generalizes the Beltrami-Laplace operator in Riemannian geometry and is defined by

$$\Delta_{\text{sub}} := \operatorname{div}_\omega \circ \operatorname{grad}_{\mathcal{H}}. \tag{2}$$

Take a *local orthonormal frame* $[X_1, \cdots, X_m]$ for $\mathcal{H}$ with $\operatorname{rank}(\mathcal{H}) = m \leq n = \dim M$. One easily checks that Δ_{sub} locally has the form of a sum-of-squares operator with a first order term depending on ω:

$$\Delta_{\text{sub}} = \sum_{i=1}^{m} \Big[X_i^2 + \operatorname{div}_\omega(X_i) \cdot X_i \Big].$$

By definition the sub-Laplacian (2) is *negative* and *symmetric* on $L^2(M, \omega)$. If M is compact, then it is *essentially self-adjoint* on $C^\infty(M)$. Different from the Laplace operator in Riemannian geometry Δ_{sub} is not elliptic in general. However, a classical theorem by Hörmander [18] together with the above local form of Δ_{sub} shows *hypoellipticity* of the sub-Laplacain. We will not give the precise definition of a hypoelliptic operator but only remark that if P is hypoelliptic, then smoothness of $f = Pu$ implies smoothness of a solution u to this equation (see [16] for related and even more general results).

Theorem 3.2 ([18]) *Let $\mathcal{L}$ be a differential operator on M, that locally in a neighborhood U of any point is written as*

$$\mathcal{L} = \sum_{i=1}^{m} Y_i^2 + Y_0,$$

where $Y_0, Y_1, \ldots, Y_m$ are C^∞-vector fields with

$$Lie_q\left\{Y_0, Y_1, \ldots, Y_m\right\} = T_qM \quad \forall\ q \in U.$$

Then $\mathcal{L}$ is hypoelliptic.

On an *equiregular SR manifold M* there is an intrinsic choice of a smooth measure ω in the definition of the sub-Laplacian. In the literature it is referred to as *Popp's measure* ω_{Popp}. Popp's measure is directly constructed from the SR structure and it generalizes the standard volume form in Riemannian geometry. Its construction essentially makes use of the bracket generating property of $\mathcal{H}$. We mention that ω_{Popp} shares some properties with the Riemannian volume form. It is known that SR isometries (as defined in [3]) preserve ω_{Popp}. If the SR isometry group acts transitively on M, then ω_{Popp} is unique with this property up to constant multiple, see [3]. If we choose $\omega = \omega_{\text{Popp}}$ in the definition (2) to be Popp's measure, then we call Δ_{sub} *intrinsic sub-Laplacian in M* (see [2, 5, 20] for details of the construction and some explicit formulas).

Example: Recall that the Martinet distribution in Example 2 is not equiregular. However, if we restrict $\mathcal{H}$ to $M_r := \{(x, y, z) \in \mathbb{R}^3 : y \neq 0\}$ and define a SR metric $\langle\cdot,\cdot\rangle$ on the restricted distribution by assuming that the vector fields X and Y in (1) are pointwise orthonormal, then $(M_r, \mathcal{H}, \langle\cdot,\cdot\rangle)$ defines an equiregular SR manifold. The intrinsic sub-Laplacian on M_r has the form:

$$\Delta_{\text{sub}} = X^2 + Y^2 - \frac{1}{y}\,Y.$$

Note that Δ_{sub} contains a non-vanishing first order term which becomes *singular* at the $\{y = 0\}$-hypersurface (boundary).

3 Subriemannian Heat Equation and Spectrum

Partial differential equations (PDE) containing the sub-Laplacian or the potential and spectral theory of the sub-Laplacian itself has been studied intensively during the last years. Interesting connections between geometry and analysis have been found, cf. [2, 8–11, 13, 18–20, 22]. In many aspects results differ from the more classical situation of the (elliptic) Laplacian on a Riemannian manifold and still a number of important questions remain open.

Let $M = (M, \mathcal{H}, \langle\cdot,\cdot\rangle)$ be an equiregular SR manifold with intrinsic sub-Laplacian Δ_{sub}. We consider the SR heat equation, which arises by formally replacing in the standard heat equation the Beltrami-Laplace operator in Riemannian geometry with the sub-Laplacian. More precisely, consider the SR heat operator

$$P := \frac{\partial}{\partial t} - \Delta_{\text{sub}} \quad \textit{on} \quad \widehat{M} := M \times \mathbb{R}_+, \quad t \in \mathbb{R}_+.$$

The SR heat equation has the form $Pu = 0$.

Definition 3.4 The SR *heat kernel* of the sub-Laplacian Δ_{sub} on M, denoted by:

$$K_{\text{sub}}(t; x, y) : (0, \infty) \times M \times M \longrightarrow \mathbb{R},$$

is the *fundamental solution* of the *SR heat operator*, i.e.

$$\begin{cases} PK_{\text{sub}}(t; \cdot, y) = 0, & \textit{for all } t > 0, \ y \in M \\ \lim_{t \downarrow 0} K_{\text{sub}}(t; x, \cdot) = \delta_x, & \textit{in the distributional sense for all } x \in M. \end{cases}$$

We will always assume that the metric space (M, d_{cc}) is complete. In this case Δ_{sub} is essentially selfadjoint on compactly supported smooth functions $C_c^\infty(M)$ and the existence and uniqueness of K_{sub} is guaranteed. Note that Theorem 3.2 applied to P acting on $\widehat{M}$ implies that P is hypoelliptic, as well. Hence, from the equation:

$$PK_{\text{sub}}(t; \cdot, y) = 0$$

and the symmetry of $K_{\text{sub}}(t; x, y)$ with respect to the space variables x and y one concludes that K_{sub} is a smooth function (not just a distribution). Analogously to the heat kernel in Riemannian geometry one may expect to find some some geometric information from the small time asymptotic expansion of the SR heat kernel. It is well-known that K_{sub} for diagonal values $(x, x) \in M \times M$ admits an asymptotic expansion of the following form (G. Ben Arous, R. Leandré): let $x \in M$:

$$K_{\text{sub}}(t, x, x) = \frac{1}{t^{\frac{Q(x)}{2}}} \Big(c_0(x) + c_1(x)t + \cdots + c_N(x)t^N + O(t^{N+1}) \Big) \quad \textit{as} \quad t \downarrow 0.$$

In our setting of an equiregular SR manifold the exponent $Q(x)$ does not depend on $x \in M$ and is known to coincide with the *Hausdorff dimension* of the metric space (M, d_{cc}). Finding the geometric content of the coefficients $c_j(x)$ $j = 1, 2 \ldots$ is a challenging problem. For some SR manifolds with additional structure $c_1(x)$ has been expressed in terms of suitable curvature terms corresponding to a connection comparable with the SR structure (see [9] in the case of so-called H-type foliations M). We can recover $d_{\text{cc}}(x, y)$ for $x, y \in M$ from off-diagonal values of K_{sub}:

Theorem 3.3 (Léandre 1987)

$$\lim_{t \downarrow 0} t \log K_{sub}(t; x, y) = -\frac{d_{cc}(x, y)^2}{2}.$$

In case of a Riemannian manifold this relation is called Varadhan formula.

Another point of view is the analysis of the *spectral zeta function* of the sub-Laplacian. Assume that M is compact. Based on subelliptic estimates for Δ_{sub} (which are a main ingredient in the proof of Theorem 3.2) it can be shown that $-\Delta_{\text{sub}}$ has discrete spectrum $\sigma\big(-\Delta_{\text{sub}}\big) = \{0 \leq \lambda_1 < \lambda_2 < \lambda_3 \cdots\}$ consisting of non-negative eigenvalues λ_j with finite multiplicities $m(\lambda_j)$. We may define the *spectral zeta function* in SR geometry in the usual way as the parameter series:

$$\zeta_{\text{sub}}(s) := \sum_{\lambda_j \neq 0} \frac{m(\lambda_j)}{\lambda_j^s}, \text{ s} \in \mathbb{C} \textit{ and } \operatorname{Re}(s) \textit{ sufficiently large.}$$

What are the relations between geometric data and the singularities of ζ_{sub}? Does it admit a meromorphic extension? If yes, how are the poles distributed and what are the residues? Does $\zeta_{\text{sub}}(s)$ extend holomorphically to $s = 0$ such that a regularized determinant gives a spectral invariant in a standard way? For certain classes of SR manifolds such questions have been addressed in the literature (e.g. see [7, 15]).

Acknowledgements The author's visit to Ghent University was supported by the Doctoral Schools of Ghent University and by the Methusalem programme of the Ghent University Special Research Fund (BOF) (Grant number 01M01021).

References

1. Agrachev, A., Barilari, D., Boscain, U.: A Comprehensive Introduction to Sub-Riemannian Geometry. Cambridge University Press, Cambridge (2019)
2. Agrachev, A., Boscain, U., Gauthier, J.-P., Rossi, F.: The intrinsic hypoelliptic Laplacian and its heat kernel on unimodular Lie groups. J. Funct. Anal. **256**, 2621–2655 (2009)
3. Barilari, D., Rizzi, L.: A formula for Popp's volume in sub-Riemannian geometry. Anal. Geom. Metric Spaces **1**, 42–57 (2013)
4. Bauer, W., Furutani, K., Iwasaki, C.: A codimension 3 sub-Riemannian structure on the Gromoll-Meyer exotic sphere. Diff. Geom. Appl. **53**, 114–136 (2017)
5. Bauer, W., Furutani, K., Iwasaki, C.: Sub-Riemannian structures in a principal bundle and their Popp measures. Appl. Anal. **96**(14), 2390–2407 (2017)
6. Bauer, W., Furutani, K., Iwasaki, C.: Trivializable sub-Riemannian structures on spheres. Bull. Sci. Math. **137**, 361–385 (2013)
7. Bauer, W., Furutani, K., Iwasaki, C.: Spectral zeta function of the sub-Laplacian on two step nilmanifolds. J. Math. Pures Appl. **97**, 242–261 (2012)
8. Bauer, W., Laaroussi, A.: Trivializable and quaternionic subriemannian structures on $\mathbb{S}^7$ and subelliptic heat kernel. J. Geom. Anal. **32**(8), 1–37 (2022)
9. Bauer, W., Markina, I., Laaroussi, A., Vega-Molino, S.: Local invariants and geometry of the sub-Laplacian on H-type foliations. Nonlinear Anal. TMA **262**(2026), 113934 (2022). arXiv: 2209.02168v1
10. Bauer, W., Laaroussi, A., Tarama, D.: Rank 5 trivializable subriemannian structure on $\mathbb{S}^7$ and subelliptic heat kernel. Potential Anal. **61**, 289–316 (2024)
11. Bonfiglioli, A., Lanconelli, F., Uguzzoni, F.: Stratified Lie Groups and Potential Theory for their Sub-Laplacians. Springer Monographs in Mathematics (2007)
12. Calin, O., Chang, D.-C.: Subriemannian Geometry—General Theory and Examples. Cambridge University Press, Cambridge (2009)

13. Calin, O., Chang, D.-C., Furutani, K., Iwasaki, C.: Heat Kernels for Elliptic and Sub-elliptic Operators: Methods and Techniques. Applied and Numerical Harmonic Analysis. Birkhäuser/Springer, New York (2011)
14. Chow, W.L.: Über Systeme von linearen partiellen Differentialgleichungen erster Ordnung. Math. Ann. **117**, 98–105 (1939)
15. Fischer, V.: Asymptotics and zeta functions on compact nilmanifolds. J. Math. Pures Appl. **160**, 1–28 (2022)
16. Fischer, V., Ruzhansky, M.: Quantization on Nilpotent Lie Groups, vol. 314. Birkhäuser, Cham, Switzerland (2016)
17. Gromov, M.: Carnot-Carathéodory spaces seen from within. In: Bellaiche, A., Risler, J.-J. (eds.) Subriemannian Geometry, pp. 79–323 (1996)
18. Hörmander, L.: Hypoelliptic second order differential operators. Acta Math. **119**, 147–171 (1967)
19. Markina, I.: Geodesics in geometry with constraints and applications. Quantization, PDEs, and geometry. In: Operator Theory: Advances and Applications. Advances in Partial Differential Equations (Basel), vol. 251, pp. 153–314. Birkhäuser/Springer, Cham (2016)
20. Montgomery, R.: A Tour of Subriemannian Geometries, their Geodesics and Applications. AMS Mathematical Surveys and Monographs, vol. 91 (2002)
21. Rashevsky, P.K.: Any two points of a totally nonholonomic space may be connected by an admissible line. Uch. Zap. Ped. Inst. im. Liebknechta, Ser. Phys. Math. **2**, 83–94 (1938)
22. Strichartz, R.: Sub-Riemannian geometry. J. Diff. Geom. **24**(2), 221–263 (1986)
23. Strichartz, R.: Corrections to "Sub-Riemannian geometry". J. Diff. Geom. **30**(2), 595–596 (1989)

Stability Results for Sobolev, Logarithmic Sobolev, and Related Inequalities

Jean Dolbeault

Abstract Obtaining explicit stability estimates in classical functional inequalities like the Sobolev inequality has been an essentially open question for 30 years, after the celebrated but non-constructive result of Bianchi and Egnell (J. Funct. Anal. 100(1):18–24, 1991, [4]). Recently, new methods have emerged which provide some clues on these fascinating questions. The goal of the course is to give an introduction to the topic for some fundamental functional inequalities and present several methods that can be used to obtain explicit estimates.

Keywords Stability · Functional inequalities · Sobolev inequality · Logarithmic sobolev inequality · Gagliardo-Nirenberg inequalities · Constructive estimates · Nonlinear diffusions · Entropy methods · Carré du champ

MSC2020 49J40 · 26D10 · 35A23

1 Introduction

The most classical Sobolev inequality on $\mathbb{R}^d$ with $d \geq 3$ is

$$\|\nabla f\|^2_{\mathrm{L}^2(\mathbb{R}^d)} \geq \mathsf{S}_d \ \|f\|^2_{\mathrm{L}^{2^*}(\mathbb{R}^d)} \quad \forall\, f \in \mathscr{D}^{1,2}(\mathbb{R}^d) \tag{1}$$

where $\mathscr{D}^{1,2}(\mathbb{R}^d)$ is the Beppo-Levi space, S_d denotes the sharp Sobolev constant and $2^* = 2\,d/(d-2)$ is the critical Sobolev exponent. Equality in (1) holds if and only if f is in the $(d+2)$-dimensional manifold $\mathscr{M}$ of the *Aubin–Talenti functions*

$$g_{a,b,c}(x) = c(a + |x-b|^2)^{-\frac{d-2}{2}}, \quad a \in (0,\infty), \quad b \in \mathbb{R}^d, \quad c \in \mathbb{R}.$$

J. Dolbeault (✉)
CEREMADE (CNRS UMR no 7534), PSL University, Université Paris-Dauphine, Paris, France
e-mail: dolbeaul@ceremade.dauphine.fr

M. Chatzakou et al. (eds.), *Direct and Inverse Problems with Applications*,
Trends in Mathematics 12, https://doi.org/10.1007/978-3-031-98645-1_4

We refer to [39] for an up-to-date introduction to Ineq. (1). In 1991, Bianchi and Egnell proved in [4] that the *deficit* in the Sobolev inequality, that is, the difference of the two sides of (1), satisfies the *stability* inequality

$$\|\nabla f\|^2_{\mathrm{L}^2(\mathbb{R}^d)} - \mathsf{S}_d\,\|f\|^2_{\mathrm{L}^{2^*}(\mathbb{R}^d)} \ge \kappa_d \min_{g\in\mathscr{M}} \|\nabla(f-g)\|^2_{\mathrm{L}^2(\mathbb{R}^d)} \tag{2}$$

for some constant $\kappa_d > 0$. The method of [4] uses compactness and provides no estimate on κ_d. Recently, several *constructive* results have been obtained, i.e., results with explicit estimates of κ_d. The goal of these lecture notes is to make a short review based on two main strategies:

1. The strategy of G. Bianchi and H. Egnell made explicit in both of its steps: the global-to-local reduction and the local analysis, both relying on methods of the Calculus of Variations,
2. *Entropy methods* and *fast diffusion equations*, in which we can recognize also the same two steps, that are based on nonlinear parabolic flows and *carré du champ* techniques.

Entropy methods usually result in better estimates of the stability constant, but to the price of some limitations. We will also state some results for related inequalities, like some families of Gagliardo-Nirenberg interpolation inequalities and logarithmic Sobolev inequalities, on the Euclidean space $\mathbb{R}^d$ or on the unit sphere $\mathbb{S}^d \subset \mathbb{R}^{d+1}$, and consider the use of the Legendre duality applied, for instance, to Hardy-Littlewood-Sobolev inequalities.

2 Sobolev and Hardy-Littlewood-Sobolev Inequalities

Let us start with a first example of fast diffusion flows applied to inequalities in the context of the Legendre duality. On $\mathbb{R}^d$ with $d \ge 3$, the *Hardy-Littlewood-Sobolev inequality*

$$\|g\|^2_{\mathrm{L}^{\frac{2d}{d+2}}(\mathbb{R}^d)} \ge \mathsf{S}_d \int_{\mathbb{R}^d} g\,(-\Delta)^{-1} g\,dx \quad \forall\, g \in \mathrm{L}^{\frac{2d}{d+2}}(\mathbb{R}^d) \tag{3}$$

is obtained by Legendre's duality from (1), as observed in [49]. Indeed, by the Legendre transform, we obtain

$$\sup_{f\in\mathscr{D}^{1,2}(\mathbb{R}^d)} \left(\int_{\mathbb{R}^d} f\,g\,dx - \frac{1}{2}\,\|f\|^2_{\mathrm{L}^{2^*}(\mathbb{R}^d)}\right) = \frac{1}{2}\,\|g\|^2_{\mathrm{L}^{\frac{2d}{d+2}}(\mathbb{R}^d)}$$

$$\sup_{f\in\mathscr{D}^{1,2}(\mathbb{R}^d)} \left(\int_{\mathbb{R}^d} f\,g\,dx - \frac{1}{2}\,\|\nabla f\|^2_{\mathrm{L}^2(\mathbb{R}^d)}\right) = \frac{1}{2}\int_{\mathbb{R}^d} g\,(-\Delta)^{-1} g\,dx$$

where $q = \frac{d+2}{d-2}$ is the Hölder conjugate of 2^*, so that $\|f^q\|_{\mathrm{L}^{\frac{2d}{d+2}}(\mathbb{R}^d)} = \|f\|^q_{\mathrm{L}^{2^*}(\mathbb{R}^d)}$. It is not difficult to check that S_d is also the optimal constant in (3).

Theorem 4.1 ([33]) *Assume that $d \ge 3$. There exists a constant $\mathscr{C} \in [\frac{d}{d+4}, 1)$ such that*

$$\begin{aligned} &\|f^q\|^2_{\mathrm{L}^{\frac{2d}{d+2}}(\mathbb{R}^d)} - \mathsf{S}_d \int_{\mathbb{R}^d} f^q\,(-\Delta)^{-1} f^q\,dx \\ &\le \mathscr{C}\,\mathsf{S}_d^{-1}\,\|f\|^{\frac{8}{d-2}}_{\mathrm{L}^{2^*}(\mathbb{R}^d)} \left(\|\nabla f\|^2_{\mathrm{L}^2(\mathbb{R}^d)} - \mathsf{S}_d\,\|f\|^2_{\mathrm{L}^{2^*}(\mathbb{R}^d)}\right) \quad \forall\, f \in \mathscr{D}^{1,2}(\mathbb{R}^d). \end{aligned} \tag{4}$$

Theorem 4.1 can be considered as a stability result for (1), because the left-hand side in (4), which is the deficit associated with (3), measures a distance to $\mathscr{M}$. However, it is still a nonlinear functional and corresponds to a weaker norm than the usual norm in $\mathscr{D}^{1,2}(\mathbb{R}^d)$. With $\mathscr{C} = 1$, there is an easy proof that goes as follows. Two integrations by parts show that

$$\int_{\mathbb{R}^d} |\nabla(-\Delta)^{-1}\,g|^2\,dx = \int_{\mathbb{R}^d} g\,(-\Delta)^{-1}\,g\,dx$$

and, if $g = f^q$ with $q = \frac{d+2}{d-2}$,

$$\int_{\mathbb{R}^d} \nabla f \cdot \nabla(-\Delta)^{-1}\,g\,dx = \int_{\mathbb{R}^d} f\,g\,dx = \int_{\mathbb{R}^d} g^q\,dx = \int_{\mathbb{R}^d} f^{2^*}\,dx.$$

Hence the expansion of the square

$$0 \le \int_{\mathbb{R}^d} \left|\,\|f\|^{\frac{4}{d-2}}_{\mathrm{L}^{2^*}(\mathbb{R}^d)}\,\nabla f - \mathsf{S}_d\,\nabla(-\Delta)^{-1}\,g\right|^2 dx$$

shows the result with $\mathscr{C} = 1$. To prove that the inequality holds with $\mathscr{C} < 1$ requires finer tools.

Let us consider the *fast diffusion* equation

$$\frac{\partial v}{\partial t} = \Delta v^m, \quad (t,x) \in \mathbb{R}^+ \times \mathbb{R}^d \tag{5}$$

and compute the evolution of the deficit in (3) defined as

$$-\,\mathsf{H} := \|v\|^2_{\mathrm{L}^{\frac{2d}{d+2}}(\mathbb{R}^d)} - \mathsf{S}_d \int_{\mathbb{R}^d} v\,(-\Delta)^{-1} v\,dx \ge 0.$$

As in [27], we observe that

$$\frac{1}{2}\,\mathsf{H}' = \left(\int_{\mathbb{R}^d} v^{\frac{2d}{d+2}}\,dx\right)^{\frac{2}{d}} \int_{\mathbb{R}^d} \nabla v^m \cdot \nabla v^{\frac{d-2}{d+2}}\,dx - \mathsf{S}_d \int_{\mathbb{R}^d} v^{m+1}\,dx$$

where $v = v(t,\cdot)$ is a solution of (5). With the choice $m = (d-2)/(d+2)$, that is, $m+1 = 2d/(d+2) = q$, which corresponds to the exponent in the *Yamabe flow*, and $u = v^m$, we have

$$\frac{1}{2}\frac{d}{dt}\left(\mathsf{S}_d \int_{\mathbb{R}^d} v\,(-\Delta)^{-1} v\,dx - \|v\|^2_{\mathrm{L}^{\frac{2d}{d+2}}(\mathbb{R}^d)}\right)$$
$$= \left(\int_{\mathbb{R}^d} v^{m+1}\,dx\right)^{\frac{2}{d}} \left(\|\nabla u\|^2_{\mathrm{L}^2(\mathbb{R}^d)} - \mathsf{S}_d\,\|u\|^2_{\mathrm{L}^{2^*}(\mathbb{R}^d)}\right) \geq 0.$$

Equation (5) admits a solution with *separation of variables* vanishing at $t = T > 0$,

$$\overline{v}_T(t,x) = c\,(T-t)^{\alpha}\,(g_\star(x))^{\frac{d+2}{d-2}}, \quad \alpha = \tfrac{1}{4}\,(d+2) \quad \text{and} \quad c = \left(4\,d\,\tfrac{d-2}{d+2}\right)^{\alpha},$$

where $g_\star(x) = (1+|x|^2)^{-(d-2)/2}$ is the solution of $\Delta g_\star + d\,(d-2)\,g_\star^{(d+2)/(d-2)} = 0$. Up to conformal transformations, this solution describes the asymptotic extinction profile of all solutions of (5) near the extinction time T.

Theorem 4.2 ([23]) For any solution v with initial datum $v_0 \in \mathrm{L}^{2d/(d+2)}(\mathbb{R}^d)$, $v_0 > 0$, there exists $T > 0$, $\lambda > 0$ and $x_0 \in \mathbb{R}^d$ such that

$$\lim_{t\to T_-} (T-t)^{-\frac{1}{1-m}} \sup_{x\in\mathbb{R}^d} (1+|x|^2)^{d+2} \left|\frac{v(t,x)}{\overline{v}(t,x)} - 1\right| = 0$$

with $\overline{v}(t,x) = \lambda^{-(d+2)/2}\,\overline{v}_T\big(t,(x-x_0)/\lambda\big)$.

The Hardy-Littlewood-Sobolev inequality (3) amounts to $\mathsf{H} \leq 0$ and appears as a consequence of the Sobolev inequality (1) because $\mathsf{H}' \geq 0$ for any $t \in (0,T)$ and $\lim_{t\to T_-} \mathsf{H}(t) = 0$ by Theorem 4.2. By computing H'', one can obtain refined estimates. Assume that $t \in [0,T)$ and let Y be such that

$$\mathsf{H}(t) = -\,\mathsf{Y}(\mathsf{J}(t)) \quad \text{with} \quad \mathsf{J}(t) := \int_{\mathbb{R}^d} v(t,x)^{\frac{2d}{d+2}}\,dx.$$

With $\kappa_0 = \mathsf{H}_0'/\mathsf{J}_0$, we obtain the differential inequality

$$\mathsf{Y}'\left(\mathsf{S}_d\, s^{1+\frac{2}{d}} + \mathsf{Y}\right) \le \tfrac{d+2}{2d}\,\kappa_0\,\mathsf{S}_d^2\, s^{1+\frac{4}{d}}, \quad \mathsf{Y}(0) = 0, \quad \mathsf{Y}(\mathsf{J}_0) = -\,\mathsf{H}_0.$$

The relation $\mathsf{Y}'(\mathsf{J})' + \mathsf{H}' = 0$ and elementary integrations by parts provide the following refinement of (4) written with $\mathscr{C} = 1$.

Proposition 4.1 ([33]) *Assume that $d \ge 3$. Then for any $f \in \mathscr{D}^{1,2}(\mathbb{R}^d)$ we have*

$$\|f^q\|^2_{\mathrm{L}^{\frac{2d}{d+2}}(\mathbb{R}^d)} - \mathsf{S}_d \int_{\mathbb{R}^d} f^q\,(-\Delta)^{-1} f^q\,dx$$

$$\le \|f\|^{2\frac{d+2}{d-2}}_{\mathrm{L}^{2^*}(\mathbb{R}^d)}\ \varphi\left(\mathsf{S}_d^{-1}\ \|f\|^{-2}_{\mathrm{L}^{2^*}(\mathbb{R}^d)}\left(\|\nabla f\|^2_{\mathrm{L}^2(\mathbb{R}^d)} - \mathsf{S}_d\ \|f\|^2_{\mathrm{L}^{2^*}(\mathbb{R}^d)}\right)\right)$$

with $q = (d+2)/(d-2)$ and $\varphi(s) := \sqrt{1+2\,s} - 1$ for any $s \ge 0$.

An asymptotic analysis of $v(t,\cdot)$ as $t \to T_-$ and some spectral estimates discard the case $\mathscr{C} = 1$ in Theorem 4.1, while the lower bound $\mathscr{C} = d/(d+4)$ is obtained by considering a sequence of test functions. See [33] for details. In dimension $d = 2$ similar results holds in which (1) and (3) are respectively replaced by Onofri's inequality and the logarithmic Hardy-Littlewood-Sobolev inequality while (5) has to be replaced by the logarithmic fast diffusion equation using the results of [50, 54, 55] instead of Theorem 4.2. Notice that stability results based on duality are known for (logarithmic) Hardy-Littlewood-Sobolev inequalities from [16, 17]. The method based on (5) also applies to the fractional Sobolev inequality and the dual Hardy-Littlewood-Sobolev inequality as shown in [45]. Even if the optimal value of the stability constant $\mathscr{C}$ is not known, a nice feature of Theorem 4.1 is that we have a rather precise estimate of $\mathscr{C}$. However, a significant drawback of the method is that the deficit of the Hardy-Littlewood-Sobolev inequality corresponds to a weaker notion of distance to $\mathscr{M}$ than what can be expected.

3 Stability, Fast Diffusion Equation and Entropy Methods

On $\mathbb{R}^d$, let us consider the *fast diffusion equation* written in self-similar variables,

$$\frac{\partial v}{\partial t} + \nabla\cdot\left(v\left(\nabla v^{m-1} - 2\,x\right)\right) = 0. \tag{6}$$

Up to a t-dependent rescaling, it is equivalent to (5). By *fast diffusion*, we mean $m < 1$ and reserve the expression *porous medium* to the regime $m > 1$ while the heat equation case corresponds to $m = 1$ (in this case, Eq. (6) is known as the *Fokker-Planck equation*, and v^{m-1} should be replaced by $-\log v$). An introduction can be found in the two classical books [52, 53]. Section 2 was devoted to $m = m_* :=$

$(d-2)/(d+2)$ with $d \geq 3$. Here we are interested in the range $m_c := (d-2)/d < m < 1$ in which nonnegative solution with initial data in $\mathrm{L}^1(\mathbb{R}^d)$ at $t=0$ exist for any $t \geq 0$ according to [42], and more specifically to the range $m_1 \leq m < 1$ where $m_1 := (d-1)/d$. Note that $m_* < m_c < m_1 < 1$ in dimension $d \geq 2$. In any case, Eq. (6) admits a stationary solution

$$\mathcal{B}(x) := \left(1+|x|^2\right)^{\frac{1}{m-1}}$$

which is usually called a *Barenblatt solution* and has finite mass $M > 0$ for any $m \in (m_c, 1)$. If $m \in (m_c, 1)$, the mass of any nonnegative solution v is conserved and for simplicity, we shall assume that $\int_{v(t,x)} = M$ for any $t \geq 0$. See [54] for an interesting survey on mass conservation and the role of Barenblatt solutions in the range $(m_c, 1)$. With this notation, we can introduce the *generalized entropy* (or *free energy*) and *Fisher information* functionals respectively given by

$$\mathcal{F}[v] := \frac{1}{m-1} \int_{\mathbb{R}^d} \left(v^m - \mathcal{B}^m - m\,\mathcal{B}^{m-1}\,(v-\mathcal{B})\right) dx,$$

$$\mathcal{I}[v] := \frac{m}{1-m} \int_{\mathbb{R}^d} v \left|\nabla v^{m-1} - \nabla \mathcal{B}^{m-1}\right|^2 dx.$$

A straightforward computation shows that, for a solution of (6), we have

$$\frac{d}{dt}\mathcal{F}[v(t,\cdot)] = -\,\mathcal{I}[v(t,\cdot)]. \tag{7}$$

Theorem 4.3 ([22]) *Assume that* $m \in (m_1, 1)$ *if* $d=1$ *or* $d=2$, *and* $m \in [m_1, 1)$ *if* $d \geq 3$. *If* v *is a nonnegative solution of (6) with initial datum* $v_0 \in \mathrm{L}^1(\mathbb{R}^d)$ *of mass* M *such that* $\mathcal{F}[v_0]$ *is finite, then*

$$\mathcal{F}[v(t,\cdot)] \leq \mathcal{F}[v_0]\,e^{-4t} \quad \forall\, t \geq 0. \tag{8}$$

The estimate (8) is equivalent to the entropy – entropy production *inequality*

$$\mathcal{I}[v] \geq 4\,\mathcal{F}[v], \tag{9}$$

which can also be rewritten as a family of Gagliardo-Nirenberg-Sobolev inequalities

$$\|\nabla f\|_{\mathrm{L}^2(\mathbb{R}^d)}^{\theta}\,\|f\|_{\mathrm{L}^{p+1}(\mathbb{R}^d)}^{1-\theta} \geq \mathscr{C}_{\mathrm{GNS}}(p)\,\|f\|_{\mathrm{L}^{2p}(\mathbb{R}^d)} \tag{10}$$

with optimal constant. The equivalence is obtained using

$$p = \frac{1}{2m-1} \quad \Longleftrightarrow \quad m = \frac{p+1}{2p}$$

with $v = f^{2p}$ *so that* $v^m = f^{p+1}$ *and* $v\,|\nabla v^{m-1}|^2 = (p-1)^2\,|\nabla f|^2$. *The range* $m \in (m_1, 1)$ *means that* $2\,p \in (2, 2^*)$ *where* $2^* = 2\,d/(d-2)$ *if* $d \geq 3$ *and* $2^* = +\infty$ *otherwise. The case* $m = m_1$, *i.e.,* $2\,p = 2^*$, *is also covered if* $d \geq 3$.

Equality in (10) is achieved by $f_\star(x) = \mathcal{B}(x)^{\frac{1}{2p}} = \big(1+|x|^2\big)^{-\frac{1}{p-1}}$, which also means that 4 is the optimal constant in (9) and it is also the best possible decay rate in (8). The exponent $\theta = \frac{d\,(p-1)}{(d+2-p\,(d-2))\,p}$ is determined by the invariance under scalings and it is easy to check that $\theta = 1$ for $p = 2^*/2$ and $d \geq 3$. If $d = 2$, we obtain the Euclidean Onofri inequality in the limit as $p \to +\infty$. On the other hand, $\lim_{p\to 1_+} \mathscr{C}_{\mathrm{GNS}}(p) = 1$ so that (10) becomes an equality for $p = 1$. By taking the derivative with respect to p at $p = 1$, we obtain the scale invariant form of the scale invariant *Euclidean logarithmic Sobolev inequality*

$$\frac{d}{2}\,\log\left(\frac{2}{\pi\,d\,e}\int_{\mathbb{R}^d}|\nabla f|^2\,dx\right) \geq \int_{\mathbb{R}^d}|f|^2\,\log|f|^2\,dx \tag{11}$$

for any function $f \in \mathrm{H}^1(\mathbb{R}^d)$ such that $\|f\|_{\mathrm{L}^2(\mathbb{R}^d)} = 1$. Ineq. (11) can be found in [56, Theorem 2], [51, Inequality (2.3)] and [15, Inequality (26)]. See [35] for further comments and [40] for the classical Gaussian form and its Euclidean counterpart.

We obtain a *linearized free energy* and a *linearized Fisher information*

$$\mathsf{F}[w] := \frac{m}{2}\int_{\mathbb{R}^d} w^2\,\mathcal{B}^{2-m}\,dx \quad\text{and}\quad \mathsf{I}[w] := m\,(1-m)\int_{\mathbb{R}^d}|\nabla w|^2\,\mathcal{B}\,dx$$

by considering $f_\varepsilon := \mathcal{B}\,(1+\varepsilon\,\mathcal{B}^{1-m}\,w)$ and taking the $O(\varepsilon^2)$ in, respectively $\mathcal{F}[f_\varepsilon]$ and $\mathcal{I}[f_\varepsilon]$. The functionals F and I are related by a *Hardy-Poincaré inequality*.

Proposition 4.2 ([6, 9, 34]) *Let* $m \in [m_1, 1)$ *if* $d \geq 3$, $m \in (1/2, 1)$ *if* $d = 2$, *and* $m \in (1/3, 1)$ *if* $d = 1$. *If* $w \in \mathrm{L}^2(\mathbb{R}^d, \mathcal{B}^{2-m}\,dx)$ *is such that* $\nabla w \in \mathrm{L}^2(\mathbb{R}^d, \mathcal{B}\,dx)$, $\int_{w\,\mathcal{B}^{2-m}} = 0$, *then*

$$\mathsf{I}[w] \geq 4\,\alpha\,\mathsf{F}[w] \tag{12}$$

with $\alpha = 1$. *Additionally, under the center of mass condition*

$$\int_{\mathbb{R}^d} x\,w\,\mathcal{B}^{2-m}\,dx = 0, \tag{13}$$

then the inequality holds with $\alpha = 2 - d\,(1-m)$.

The value of α in Proposition 4.2 is optimal and corresponds to the spectral gap of the linearized evolution operator $\mathcal{L}$ associated with (6), and to an improved spectral gap under Condition (13) in the second case. In the first case, we recover that $\inf \mathcal{I}/\mathcal{F} = \inf \mathsf{I}/\mathsf{F} = 4$, which means that the worst decay rate is achieved in (8) in the *asymptotic regime* as $t \to +\infty$. The eigenspace of $\mathcal{L}$ associated with $\alpha = 1$

corresponds to the equality case in (12) and determines the spectral gap. See [26] for a description of the spectrum of $\mathcal{L}$ and [7] for an equivalence with the setting of Proposition 4.2. Under Condition (13), an improved entropy – entropy production inequality is obtained, at least for functions which are close enough to the manifold of the Barenblatt functions.

Corollary 4.1 ([7, 9, 34]) *Let $m \in (m_1, 1)$ if $d \ge 2$, $m \in (1/3, 1)$ if $d = 1$, $\eta = 2\,(d\,m - d + 1)$ and $\chi = m/(266 + 56\,m)$. If $\int_v = \mathrm{M}$, $\int_{x\,v} = 0$ and*

$$(1 - \varepsilon)\,\mathcal{B} \le v \le (1 + \varepsilon)\,\mathcal{B} \tag{14}$$

for some $\varepsilon \in (0, \chi\,\eta)$, then

$$\mathcal{I}[v] \ge (4 + \eta)\,\mathcal{F}[v].$$

If v solves (6) with $m \in (m_1, 1)$ and an initial datum v_0 with centred mass, i.e., such that $\int_{x\,v_0(x)} = 0$, then $\int_{x\,v(t,x)} = 0$ for any $t \ge 0$ so that $w = (v - \mathcal{B})\,\mathcal{B}^{m-2}$ satisfies (13). If additionally v_0 satisfies (14), then

$$\mathcal{F}[v(t,\cdot)] \le \mathcal{F}[v_0]\,e^{-(4+\eta)\,t} \quad \forall\,t \ge 0.$$

We have obtained an improved entropy – entropy production inequality and an improved decay rate of $\mathcal{F}[v(t,\cdot)]$ by comparing the nonlinear free energy $\mathcal{F}$ and the Fisher information $\mathcal{I}$ with their linearized counterparts F and I, and by getting rid of the eigenspace corresponding to $\alpha = 1$ in (12). The result of Corollary 4.1 can be extended to $m \notin [m_1, 1)$: see [6]. Under technical conditions, better rates can be achieved in some cases as in [25]. One of the issues is that Condition (14) is rather restrictive and determines only a small neighbourhood of the set of the Barenblatt functions. Here we are going to take advantage of a property of *relative uniform convergence* that goes back to [11]: Condition (14) is satisfied by any solution $v(t,\cdot)$ for $t \ge t_\star$ if the *threshold time* $t_\star$ is taken large enough. Based on results of [9, 10], we have the following result.

Theorem 4.4 ([8, 9]) *Assume that $m \in (m_1, 1)$ if $d \ge 2$, $m \in (1/3, 1)$ if $d = 1$ and let $\varepsilon \in (0, 1/2)$, small enough, and $A > 0$ be given. There exists an explicit threshold time $t_\star \ge 0$ such that, if v is a solution of (6) with nonnegative initial datum $v_0 \in \mathrm{L}^1(\mathbb{R}^d)$ satisfying $\int_{v_0} = \int_{\mathcal{B}} = \mathrm{M}$ and*

$$\mathsf{A}[v_0] = \sup_{r>0} r^{\frac{d\,(m-m_c)}{(1-m)}} \int_{|x|>r} v_0\,dx \le A < \infty, \tag{15}$$

and if $B(t,\cdot)$ solves (5) with initial datum $\mathcal{B}$, then

$$\sup_{x\in\mathbb{R}^d} \left| \frac{v(t,x)}{\mathcal{B}(x)} - 1 \right| \le \varepsilon \quad \forall\,t \ge t_\star.$$

Under Condition (15), we have an improved decay rate for any t in the *asymptotic time range* $[t_\star, +\infty)$.

The next question is to find an improvement in the *initial time layer* $[0, t_\star)$ as well. A key ingredient is the *carré du champ* method for which we can refer to [9, Chap. 2]. We learn for instance that for some strictly convex function ψ with $\psi(0) = 0$, $\psi'(0) = 0$, there is an improved entropy – entropy production inequality

$$\mathcal{I} - 4\mathcal{F} \geq \psi(\mathcal{F}) \geq 0,$$

so that better decay rates can be expected if $\mathcal{F}$ is large, which is precisely what can be expected in the initial time layer. In fact, a key estimate of the *carré du champ* method is the simpler observation that

$$\frac{d}{dt}\mathcal{I}[v(t,\cdot)] \leq -4\,\mathcal{I}[v(t,\cdot)]$$

if v solves (6), which allows us to prove a *backward in time* estimate. Let us define $\mathcal{Q}[v] := \frac{\mathcal{I}[v]}{\mathcal{F}[v]}$ and notice that, using (7), we have

$$\frac{d}{dt}\mathcal{Q}[v(t,\cdot)] \leq \mathcal{Q}[v(t,\cdot)]\,\big(\mathcal{Q}[v(t,\cdot)] - 4\big).$$

Lemma 4.1 ([9]) *Assume that $m > m_1$ and v is a solution to (6) with nonnegative initial datum v_0. If for some $\eta > 0$ and $t_\star \geq 0$, we have $\mathcal{Q}[v(t_\star,\cdot)] \geq 4 + \eta$, then*

$$\mathcal{Q}[v(t,\cdot)] \geq 4 + \zeta \quad \forall\, t \in [0, t_\star] \quad \textit{with} \quad \zeta := \frac{4\,\eta\, e^{-4\,t_\star}}{4 + \eta - \eta\, e^{-4\,t_\star}}.$$

In the subcritical range with $m > m_1$, the first consequence of Lemma 4.1 is an improved decay rate of the free energy for the solution of (6).

Corollary 4.2 ([9]) *Let $m \in (m_1, 1)$ if $d \geq 2$, $m \in (1/2, 1)$ if $d = 1$, and $A > 0$. If v is a solution of (6) with nonnegative initial datum $v_0 \in \mathrm{L}^1(\mathbb{R}^d)$ such that $\int_{\mathbb{R}^d} v_0\,dx = M$, $\int_{\mathbb{R}^d} x\, v_0\,dx = 0$ and v_0 satisfies (15), then there is an explicit $\zeta > 0$ such that*

$$\mathcal{F}[v(t,.)] \leq \mathcal{F}[v_0]\, e^{-(4+\zeta)\,t} \quad \forall\, t \geq 0.$$

The constant ζ can be explicitly computed in terms of d, m and A using Corollary 4.1, Theorem 4.4, and Lemma 4.1. The second consequence of Lemma 4.1 is a *stability result for the entropy – entropy production inequality*, which reads as

$$\mathcal{I}[v] - 4\mathcal{F}[v] \geq \frac{\zeta}{4+\zeta}\,\mathcal{I}[v]$$

where the left-hand side is the deficit in (9), while the distance in the right-hand side is measured by the Fisher information, which is a quantity that controls strong norms.

In the critical case $m = m_1$ corresponding to the Sobolev inequality for $d \ge 3$, we also obtain a *constructive stability result*. Let $\mathsf{g}(x) := (1 + |x|^2)^{-(d-2)/2}$ be the Aubin–Talenti function such that $\mathsf{g}^{2^*} = \mathcal{B}$.

Theorem 4.5 ([9]) *Let $d \ge 3$, $A > 0$, and $p = 2^*/2$. Then for any nonnegative $f \in \mathrm{L}^{p+1}(\mathbb{R}^d)$ such that $\nabla f \in \mathrm{L}^2(\mathbb{R}^d)$ and $|x|\, f^p \in \mathrm{L}^2(\mathbb{R}^d)$,*

$$\int_{\mathbb{R}^d} (1, x, |x|^2)\, f^{2^*}\, dx = \int_{\mathbb{R}^d} (1, x, |x|^2)\, \mathsf{g}\, dx \quad \text{and} \quad \sup_{r>0} r^d \int_{|x|>r} f^{2^*}\, dx \le A,$$

then we have

$$\|\nabla f\|^2_{\mathrm{L}^2(\mathbb{R}^d)} - \mathsf{S}_d^2\, \|f\|^2_{\mathrm{L}^{2^*}(\mathbb{R}^d)} \ge \mathscr{C} \int_{\mathbb{R}^d} \left|(p-1)\, \nabla f + f^p\, \nabla \mathsf{g}^{1-p}\right|^2 dx$$

for some explicit $\mathscr{C} > 0$ which depends only on d and A.

Extending the subcritical result of Corollary 4.2 to the critical case uses the introduction of a time delay, as shown in [9, Chap. 6]. One can also obtain a stability result in the subcritical range $m \in (m_1, 1)$, that is, $1 < p < 2^*/2$ for Ineq. (10), but the statement is more delicate as several scales are involved. A remarkable feature in Theorem 4.5 is that we measure the distance to a specific Aubin–Talenti function and not to the manifold $\mathscr{M}$.

Summing up, we have obtained a construction of an explicit stability result using the *fast diffusion equation* and *entropy methods* based on a parabolic regularity theory made explicit, spectral estimates and various features of the *carré du champ* method. However, estimates of the stability constants are way smaller than the expected values of the optimal stability constants and are available only under the restriction (15) on the tails of the functions, which is inherent to the method, as one can prove that $t_\star$ diverges if $A \to +\infty$ in (15).

4 Explicit Stability Results for Sobolev and Log-Sobolev Inequalities, with Optimal Dimensional Dependence

In this section, we just state the main results and refer the reader to [28] for detailed results and to [29] for a shorter introduction to the main tools. The most important fact is that the method gives an explicit value to the constant in the *stability* inequality (2), but again one can suspect that the lower bound is not very good, even if it has an optimal dimensional dependence.

Theorem 4.6 ([28]) *There is a constant $\beta > 0$ with an explicit lower estimate which does not depend on d such that for all $d \geq 3$ and all $f \in \mathrm{H}^1(\mathbb{R}^d)$ we have*

$$\|\nabla f\|^2_{\mathrm{L}^2(\mathbb{R}^d)} - \mathsf{S}_d \, \|f\|^2_{\mathrm{L}^{2^*}(\mathbb{R}^d)} \geq \frac{\beta}{d} \inf_{g \in \mathscr{M}} \|\nabla f - \nabla g\|^2_{\mathrm{L}^2(\mathbb{R}^d)} \,. \tag{16}$$

Using the inverse stereographic projection we can rewrite the result on $\mathbb{S}^d$ as

$$\begin{aligned}
&\|\nabla F\|^2_{\mathrm{L}^2(\mathbb{S}^d)} - \tfrac{1}{4}\, d\,(d-2) \left(\|F\|^2_{\mathrm{L}^{2^*}(\mathbb{S}^d)} - \|F\|^2_{\mathrm{L}^2(\mathbb{S}^d)}\right) \\
&\geq \frac{\beta}{d} \inf_{G \in \mathscr{M}(\mathbb{S}^d)} \left(\|\nabla F - \nabla G\|^2_{\mathrm{L}^2(\mathbb{S}^d)} + \tfrac{1}{4}\, d\,(d-2)\, \|F - G\|^2_{\mathrm{L}^2(\mathbb{S}^d)}\right).
\end{aligned}$$

Here we take on $\mathbb{S}^d$ the uniform probability measure. Let $v \in \mathrm{H}^1(\mathbb{R}^n, dx)$ with compact support, $d \geq n$ large enough and consider

$$u_d(\omega) = v\Big(\tfrac{1}{r_d}\,\omega_1, \tfrac{1}{r_d}\,\omega_2, \ldots, \tfrac{1}{r_d}\,\omega_n\Big), \quad r_d = \sqrt{\tfrac{d}{2\pi}}$$

where $\omega = (\omega_1, \omega_2, \ldots, \omega_{d+1}) \in \mathbb{S}^d \subset \mathbb{R}^{d+1}$. If $d\gamma = e^{-\pi\,|x|^2}\,dx$ denotes the Gaussian measure on $\mathbb{R}^n$, by taking the limit as $d \to +\infty$, we obtain

$$\begin{aligned}
&\lim_{d \to +\infty} d\left(\|\nabla u_d\|^2_{\mathrm{L}^2(\mathbb{S}^d)} - \tfrac{1}{4}\, d\,(d-2)\left(\|u_d\|_{\mathrm{L}^p(\mathbb{R}^d)} - \|u_d\|_{\mathrm{L}^2(\mathbb{R}^d)}\right)\right) \\
&= \|\nabla u\|^2_{\mathrm{L}^2(\mathbb{R}^n, d\gamma)} - \pi \int_{\mathbb{R}^n} |u|^2 \log\left(\frac{|u|^2}{\|u\|^2_{\mathrm{L}^2(\mathbb{R}^n, d\gamma)}}\right) d\gamma,
\end{aligned}$$

where the right-hand side is the deficit in the *Gaussian logarithmic Sobolev inequality* of [40], which is equivalent to (11). We can also let $d \to +\infty$ in (16). The delicate part is to bound the coefficients (a, b, c) of the Aubin–Talenti functions which minimize the distance in (2) uniformly with respect to d. Because of the rescaling, the gradient term in the distance is lost and stability is measured only by a $\mathrm{L}^2(\mathbb{R}^n, d\gamma)$ norm. See [28] for a detailed proof.

Corollary 4.3 ([28]) *With $\beta > 0$ as in Theorem 4.6, we have*

$$\begin{aligned}
&\|\nabla u\|^2_{\mathrm{L}^2(\mathbb{R}^n, d\gamma)} - \pi \int_{\mathbb{R}^n} |u|^2 \log\left(\frac{|u|^2}{\|u\|^2_{\mathrm{L}^2(\mathbb{R}^n, d\gamma)}}\right) d\gamma \\
&\geq \frac{\beta\,\pi}{2} \inf_{a \in \mathbb{R}^d,\, c \in \mathbb{R}} \int_{\mathbb{R}^n} |u - c\, e^{a \cdot x}|^2 \, d\gamma.
\end{aligned}$$

The proof of Theorem 4.6 relies on three main steps.

1. A sequence of nonnegative functions built using the competing symmetries of Carlen and Loss [18], which alternate a Schwarz symmetric decreasing rearrangement and a transformation which corresponds, up to the stereographic projection, to a rotation on the sphere. This step reduces the stability problem to a local problem, on a neighbourhood of $\mathscr{M}$. We also use a continuous Steiner symmetrization in the discussion of one of the cases and an argument of [21].
2. A Taylor expansion with explicitly controlled remainder terms. To obtain the d-dependence, one has to perform a delicate analysis which involves an argument of [36].
3. The result for sign-changing functions is deduced from the result for the positive and negative parts by a convexity estimate.

For full details on the proof of Theorem 4.6, on has to refer to [28]. How much is lost in the first two steps is unclear, and this is why a proof based, e.g., on a fast diffusion equation would be a significant added value. There are several results which complement Theorem 4.6 and Corollary 4.3. We know from [46] that $\beta < 4\,d/(d+4)$. The equality case in (2) written for the optimal value of κ_d is achieved according to [47]. A stability result for the Hardy-Littlewood-Sobolev inequality follows from the duality method of [16] and has been generalized in [20] to cover various exponents. This result also proves an explicit stability estimate for the fractional Sobolev inequality, which uses duality but otherwise follows the very same lines of proof as in [28].

The case of the logarithmic Hardy-Littlewood-Sobolev inequality is also obtained in [17] using duality with the Onofri inequality, whose explicit stability has been obtained in [41] by other methods.

5 More Explicit Stability Results for the Logarithmic Sobolev and Gagliardo-Nirenberg Inequalities on $\mathbb{S}^d$

Here the goal is to reconciliate the flow methods of Sect. 2 with stability estimates without restrictions as in Sect. 3 and get estimate which may have the right order of magnitude compared to the optimal stability constants. Let us consider the family of *subcritical Gagliardo-Nirenberg inequalities*

$$\|\nabla F\|^2_{\mathrm{L}^2(\mathbb{S}^d)} \ge d\,\mathcal{E}_p[F] := \frac{d}{p-2}\left(\|F\|^2_{\mathrm{L}^p(\mathbb{S}^d)} - \|F\|^2_{\mathrm{L}^2(\mathbb{S}^d)}\right) \quad \forall\, F \in \mathrm{H}^1(\mathbb{S}^d, d\mu) \tag{17}$$

for any $p \in [1,2) \cup (2,2^*)$, with $2^* := \frac{2d}{d-2}$ if $d \ge 3$ and $2^* = +\infty$ if $d = 1$ or 2. If $d \ge 3$, the limit case $p = 2^*$ of (17) is the Sobolev inequality on the sphere. As another limit case when $p \to 2$, we obtain the *logarithmic Sobolev inequality on the sphere*

$$\int_{\mathbb{S}^d} |\nabla F|^2 \, d\mu \geq d\, \mathcal{E}_2[F] := \frac{d}{2} \int_{\mathbb{S}^d} F^2 \log\left(\frac{F^2}{\|F\|^2_{\mathrm{L}^2(\mathbb{S}^d)}}\right) d\mu \quad \forall\, F \in \mathrm{H}^1(\mathbb{S}^d, d\mu) \tag{18}$$

where $d\mu$ is the uniform probability on $\mathbb{S}^d$. These inequalities were obtained in [1, 2, 5]. While equality is always achieved by constant functions, the optimal constant d can be identified with the first positive eigenvalue of the Laplace-Beltrami operator on $\mathbb{S}^d$. Let X_1 be the corresponding eigenspace of spherical harmonic functions and denote by Π_1 the orthogonal projection onto X_1. If $\varphi \in X_1$ and $F_\varepsilon = 1 + \varepsilon\, \varphi$, a simple Taylor expansion shows that $\|\nabla F_\varepsilon\|^2_{\mathrm{L}^2(\mathbb{S}^d)} - d\, \mathcal{E}_p[F_\varepsilon] = O(\varepsilon^4)$, so that there is no hope to control the deficit from below by the square of a distance based on usual norm, at least in the direction of X_1.

Theorem 4.7 ([12]) *Let $d \geq 1$ and $p \in (1, 2^*)$. There is an explicit stability constant $\mathscr{S}_{d,p} > 0$ such that, for any $F \in \mathrm{H}^1(\mathbb{S}^d, d\mu)$, we have*

$$\int_{\mathbb{S}^d} |\nabla F|^2 \, d\mu - d\, \mathcal{E}_p[F] \geq \mathscr{S}_{d,p} \left(\frac{\|\nabla \Pi_1 F\|^4_{\mathrm{L}^2(\mathbb{S}^d)}}{\|\nabla F\|^2_{\mathrm{L}^2(\mathbb{S}^d)} + \|F\|^2_{\mathrm{L}^2(\mathbb{S}^d)}} + \|\nabla(\mathrm{Id} - \Pi_1)\, F\|^2_{\mathrm{L}^2(\mathbb{S}^d)} \right).$$

This results is a consequence of the stability analysis of [38], of explicit improvements of (17) and (18) obtained for instance in [30, 32] based on [24, 31], which involve a generalized *carré du champ* method for the flow

$$\frac{\partial u}{\partial t} = u^{-p\,(1-m)} \left(\Delta u + (m\, p - 1)\, \tfrac{|\nabla u|^2}{u} \right)$$

with exponents $m \in \big[m_-(d, p), m_+(d, p)\big]$ where

$$m_\pm(d, p) := \frac{1}{(d+2)\, p} \left(d\, p + 2 \pm \sqrt{d\,(p-1)\,\big(2\,d - (d-2)\,p\big)} \right),$$

and of spectral properties taken from the proof of [3]. As in Sect. 4 and with the same notation, we can consider the large dimensional limit.

Theorem 4.8 ([13]) *Let $n \geq 1$ and $u \in \mathrm{H}^1(\mathbb{R}^n, dx)$ with compact support. For any $p \in (1, 2)$, we have*

$$\lim_{d \to +\infty} d \left(\|\nabla u_d\|^2_{\mathrm{L}^2(\mathbb{S}^d)} - \tfrac{d}{2-p} \left(\|u_d\|^2_{\mathrm{L}^2(\mathbb{S}^d)} - \|u_d\|^2_{\mathrm{L}^p(\mathbb{S}^d)} \right) \right) = \|\nabla u\|^2_{\mathrm{L}^2(\mathbb{R}^n, d\gamma)} - \tfrac{1}{2-p} \left(\|u\|^2_{\mathrm{L}^2(\mathbb{R}^n, d\gamma)} - \|u\|^2_{\mathrm{L}^p(\mathbb{R}^n, d\gamma)} \right).$$

In Theorem 4.7 we have the satisfaction to obtain a stability result which distinguishes the X_1 eigenspace, is based on a fast diffusion equation and does not involve any restriction on the function space, but the estimate is not uniform as $p \to 2^*$, as could be expected from Theorem 4.6.

There is still much more to understand, particularly concerning (18). In Corollary 4.3, the stability of the *Gaussian logarithmic Sobolev inequality* is measured only by a $\mathrm{L}^2(\mathbb{R}^n, d\gamma)$ norm. It is known from [43, 44] that an $\mathrm{H}^1(\mathbb{R}^n, d\gamma)$ stability result cannot be expected without additional conditions. The quest for optimal conditions is open but as an illustration, we can conclude with a last result.

Theorem 4.9 ([14]) *Let $d \ge 1$. For any $\varepsilon > 0$, there is some explicit $\mathscr{C} > 1$ depending only on ε such that, for any $u \in \mathrm{H}^1(\mathbb{R}^d, d\gamma)$ with*

$$\int_{\mathbb{R}^d} (1, x)\, |u|^2\, d\gamma = (1, 0), \quad \int_{\mathbb{R}^d} |u|^2\, e^{\varepsilon\,|x|^2}\, d\gamma < \infty$$

for some $\varepsilon > 0$, then we have

$$\|\nabla u\|^2_{\mathrm{L}^2(\mathbb{R}^d, d\gamma)} \ge \frac{\mathscr{C}}{2} \int_{\mathbb{R}^d} |u|^2\, \log |u|^2\, d\gamma .$$

Moreover, if u is compactly supported in a ball of radius $R > 0$, we know that the inequality holds that with $\mathscr{C} = 1 + \frac{\mathscr{C}_\star(\mathsf{K}_\star) - 1}{1 + R^2\, \mathscr{C}_\star(\mathsf{K}_\star)}$, $\mathscr{C}_\star(\mathsf{K}) = 1 + \frac{1}{432\,\mathsf{K}}$ and $\mathsf{K}_\star := \max\big(d, \frac{(d+1)\,R^2}{1+R^2}\big)$.

Here the result is stated in the form of an improved inequality under orthogonality constraints and we use the heat flow case corresponding to the limit $m = 1$ of the fast diffusion flow. The proof is based on a stability result of [37] for a restricted class of functions, which contains the solution of the heat flow after an initial time layer recently determined in [19]. For a compactly supported initial datum, we obtain a very explicit expression using a result of [48]. One can certainly object that $1/432 \approx 0.00231481$ does not provide us with a glorious improvement of the constant, but this is definitely a step in the understanding of *explicit* stability results and, as a consequence, better estimates of decay rates in nonlinear evolution problems.

Acknowledgements This work has been partially supported by the Project *Conviviality* (ANR-23-CE40-0003) of the French National Research Agency.

References

1. Bakry, D., Émery, M.: Diffusions hypercontractives. In: Séminaire de probabilités, XIX, 1983/84, vol. 1123. Lecture Notes in Mathematics, pp. 177–206. Springer, Berlin (1985)
2. Beckner, W.: Sharp sobolev inequalities on the sphere and the Moser-Trudinger inequality. Ann. Math. **138**, 213 (1993)

3. Beckner, W.: Sharp Sobolev inequalities on the sphere and the Moser-Trudinger inequality. Ann. Math. **138**, 213 (1993)
4. Bianchi, G., Egnell, H.: A note on the Sobolev inequality. J. Funct. Anal. **100**, 18–24 (1991)
5. Bidaut-Véron, M.-F., Véron, L.: Nonlinear elliptic equations on compact Riemannian manifolds and asymptotics of Emden equations. Invent. Math. **106**, 489–539 (1991)
6. Blanchet, A., Bonforte, M., Dolbeault, J., Grillo, G., Vázquez, J.L.: Asymptotics of the fast diffusion equation via entropy estimates. Arch. Ration. Mech. Anal. **191**, 347–385 (2009)
7. Bonforte, M., Dolbeault, J., Grillo, G., Vázquez, J.L.: Sharp rates of decay of solutions to the nonlinear fast diffusion equation via functional inequalities. Proc. Nat. Acad. Sci. USA **107**, 16459–16464 (2010)
8. Bonforte, M., Dolbeault, J., Nazaret, B., Simonov, N.: Constructive stability results in interpolation inequalities and explicit improvements of decay rates of fast diffusion equations. Discrete Contin. Dyn. Syst. **43**, 1070–1089 (2023)
9. Bonforte, M., Dolbeault, J., Nazaret, B., Simonov, N.: Stability in Gagliardo-Nirenberg-Sobolev inequalities: flows, regularity, and the entropy method. Mem. Amer. Math. Soc. **308**(1554):viii+166 (2025)
10. Bonforte, M., Simonov, N.: Quantitative a priori estimates for fast diffusion equations with Caffarelli-Kohn-Nirenberg weights. Harnack inequalities and Hölder continuity. Adv. Math. **345**, 1075–1161 (2019)
11. Bonforte, M., Vázquez, J.L.: Global positivity estimates and Harnack inequalities for the fast diffusion equation. J. Funct. Anal. **240**, 399–428 (2006)
12. Brigati, G., Dolbeault, J., Simonov, N.: Logarithmic Sobolev and interpolation inequalities on the sphere: constructive stability results, Ann. Inst. H. Poincaré C Anal. Non Linéaire **41**, 1289–1321 (2023)
13. Brigati, G., Dolbeault, J., Simonov, N.: On Gaussian interpolation inequalities. C. R. Math. Acad. Sci. Paris **362**, 21–44 (2024)
14. Brigati, G., Dolbeault, J., Simonov, N.: Stability for the logarithmic Sobolev inequality. J. Funct. Anal. **287**, 110562 (2024)
15. Carlen, E.A.: Superadditivity of Fisher's information and logarithmic Sobolev inequalities. J. Funct. Anal. **101**, 194–211 (1991)
16. Carlen, E.A.: Duality and stability for functional inequalities, Ann. Fac. Sci. Toulouse Math. **26**(6), 319–350 (2017)
17. Carlen, E. A.: Stability for the logarithmic Hardy-Littlewood-Sobolev inequality with application to the Keller-Segel equation. J. Funct. Anal. **288**(6), 27 (2025), Paper No. 110818
18. Carlen, E.A., Loss, M.: Extremals of functionals with competing symmetries. J. Funct. Anal. **88**, 437–456 (1990)
19. Chen, H.-B., Chewi, S., Niles-Weed, J.: Dimension-free log-Sobolev inequalities for mixture distributions. J. Funct. Anal. **281**, 109236 (2021)
20. Chen, L., Lu, G., Tang, H.: Stability of Hardy-Littlewood-Sobolev inequalities with explicit lower bounds. Adv. Math. **450**, Paper No. 109778, 28 (2024)
21. Christ, M.: A sharpened Riesz-Sobolev inequality (2017). arXiv: 1706.02007
22. Del Pino, M., Dolbeault, J.: Best constants for Gagliardo-Nirenberg inequalities and applications to nonlinear diffusions. J. Math. Pures Appl. **81**(9), 847–875 (2002)
23. del Pino, M., Sáez, M.: On the extinction profile for solutions of $u_t = \delta u^{(n-2)/(n+2)}$. Indiana Univ. Math. J. **50**, 611–628 (2001)
24. Demange, J.: Improved Gagliardo-Nirenberg-Sobolev inequalities on manifolds with positive curvature. J. Funct. Anal. **254**, 593–611 (2008)
25. Denzler, J., Koch, H., McCann, R.J.: Higher-order time asymptotics of fast diffusion in Euclidean space: a dynamical systems approach. Mem. Amer. Math. Soc. **234**, vi+81 (2015)
26. Denzler, J., McCann, R.J.: Fast diffusion to self-similarity: complete spectrum, long-time asymptotics, and numerology. Arch. Ration. Mech. Anal. **175**, 301–342 (2005)
27. Dolbeault, J.: Sobolev and Hardy-Littlewood-Sobolev inequalities: duality and fast diffusion. Math. Res. Lett. **18**, 1037–1050 (2011)

28. Dolbeault, J., Esteban, M.J., Figalli, A., Frank, R.L., Loss, M.: Sharp stability for Sobolev and log-Sobolev inequalities, with optimal dimensional dependence. Camb. J. Math. **13**(2):359–430 (2025)
29. Dolbeault, J., Esteban, M.J., Figalli, A., Frank, R.L., Loss, M.: A short review on improvements and stability for some interpolation inequalities. In: Proceedings of ICIAM 2023. Preprint arXiv:2402.08527, hal-04456461, V. S. Georgiev, T. Ozawa (eds.), Various Methods for the Analysis of PDEs, ICIAM2023. Springer Series 5. https://doi.org/10.1007/978-981-96-9164-7_3
30. Dolbeault, J., Esteban, M.J., Kowalczyk, M., Loss, M.: Improved interpolation inequalities on the sphere. Discrete Contin. Dyn. Syst. Ser. S **7**, 695–724 (2014)
31. Dolbeault, J., Esteban, M.J., Loss, M.: Nonlinear flows and rigidity results on compact manifolds. J. Funct. Anal. **267**, 1338–1363 (2014)
32. Dolbeault, J., Esteban, M.J., Loss, M.: Interpolation inequalities on the sphere: linear vs. nonlinear flows (inégalités d'interpolation sur la sphère: flots non-linéaires vs. flots linéaires). Ann. Fac. Sci. Toulouse Math. **26**(6), 351–379 (2017)
33. Dolbeault, J., Jankowiak, G.: Sobolev and Hardy-Littlewood-Sobolev inequalities. J. Diff. Eqn. **257**, 1689–1720 (2014)
34. Dolbeault, J., Toscani, G.: Fast diffusion equations: matching large time asymptotics by relative entropy methods. Kinet. Relat. Models **4**, 701–716 (2011)
35. Dolbeault, J., Toscani, G.: Nonlinear diffusions: extremal properties of Barenblatt profiles, best matching and delays. Nonlinear Anal. **138**, 31–43 (2016)
36. Duoandikoetxea, J.: Reverse Hölder inequalities for spherical harmonics. Proc. Amer. Math. Soc. **101**, 487–491 (1987)
37. Fathi, M., Indrei, E., Ledoux, M.: Quantitative logarithmic Sobolev inequalities and stability estimates. Discrete Contin. Dyn. Syst. **36**, 6835–6853 (2016)
38. Frank, R.L.: Degenerate stability of some Sobolev inequalities. Ann. Inst. H. Poincaré C Anal. Non Linéaire **39**, 1459–1484 (2022)
39. Frank, R.L.: The Sharp Sobolev Inequality and Its Stability: An Introduction. Lecture Notes in Mathematics, C.I.M.E. Foundation Subseries, vol. 2348. Springer
40. Gross, L.: Logarithmic Sobolev inequalities. Amer. J. Math. **97**, 1061–1083 (1975)
41. Gui, C., Moradifam, A.: The sphere covering inequality and its applications. Invent. Math. **214**, 1169–1204 (2018)
42. Herrero, M.A., Pierre, M.: The Cauchy problem for $u_t = \delta u^m$ when $0<m<1$. Trans. Amer. Math. Soc. **291**, 145–158 (1985)
43. Indrei, E.: W^1, 1 stability for the LSI. J. Differential Equations **421**, 196–207 (2025)
44. Indrei, E., Kim, D.: Deficit estimates for the logarithmic Sobolev inequality. Diff. Integral Eqn. **34**, 437–466 (2021)
45. Jankowiak, G., Nguyen, V.H.: Fractional Sobolev and Hardy-Littlewood-Sobolev inequalities (2014). Preprint arXiv: 1404.1028
46. König, T.: On the sharp constant in the Bianchi-Egnell stability inequality. Bull. Lond. Math. Soc. **55**, 2070–2075 (2023)
47. König, T.: Stability for the Sobolev inequality: existence of a minimizer. J. Eur. Math. Soc. (to appear). Preprint arXiv:2211.14185
48. Lee, K.-A., L. Vázquez, J.: Geometrical properties of solutions of the porous medium equation for large times. Indiana Univ. Math. J., 991–1016 (2003)
49. Lieb, E.H.: Sharp constants in the Hardy-Littlewood-Sobolev and related inequalities. Ann. Math. **118**(2), 349–374 (1983)
50. Rodriguez, A., Vázquez, J.L., Esteban, J.R.: The maximal solution of the logarithmic fast diffusion equation in two space dimensions. Adv. Diff. Eqn. **2**, 867–894 (1997)
51. Stam, A.J.: Some inequalities satisfied by the quantities of information of Fisher and Shannon. Inf. Control **2**, 101–112 (1959)
52. Vázquez, J.L.: The porous medium equation. New contractivity results. In: Elliptic and Parabolic Problems, Volume 63 of Progress in Nonlinear Differential Equations and Their Applications, Birkhäuser, Basel, pp. 433–451 (2005)

53. Vázquez, J.L.: Smoothing and Decay Estimates for Nonlinear Diffusion Equations. Volume 33 of Oxford Lecture Series in Mathematics and its Applications, Oxford University Press, Oxford (2006). Equations of porous medium type
54. Vázquez, J.L.: A survey on mass conservation and related topics in nonlinear diffusion (2023). Preprint arXiv: 2311.18357
55. Vázquez, J.L., Esteban, J.R., Rodriguez, A.: The fast diffusion equation with logarithmic nonlinearity and the evolution of conformal metrics in the plane. Adv. Diff. Eqn. **1**, 21–50 (1996)
56. Weissler, F.B.: Logarithmic Sobolev inequalities for the heat-diffusion semigroup. Trans. Amer. Math. Soc. **237**, 255–269 (1978)

Semiconcavity, Viscosity Solutions and the Square Distance in Carnot Groups

Federica Dragoni

Abstract We give an overview of semiconcavity, starting from the standard notion up to more recent generalizations in a different geometrical context, such as Carnot groups, focusing in particular on the viscosity characterization by bounds for second derivatives. We then apply these theories to show some recent results obtained by the author, in collaboration with Qing Liu and Ye Zhang. In particular, we show that the square Carnot-Carathéodory distance is h-semiconcave in step 2 Carnot groups.

Keywords Carnot groups · Heisenberg group · Convexity · Semiconcavity · Sub-Riemannian geometries · Hömander condition · Hamilton-Jacobi equations

MSC2020 30L99 · 58J65 · 31E05

1 The Euclidean Case

In this section, we briefly recall the standard notions of concavity and semiconcavity used in the Euclidean space. We also give a quick overview of viscosity solutions, focusing in particular on the viscosity characterization for concave and semiconcave functions by bounds for the second derivatives. The theory is usually stated for convex and semiconvex functions, but since the focus later is the semiconcavity of the square distance, we state all definitions and properties for concave and semiconcave functions. We recall that a function u is concave (respectively, semiconcave) if and only if $-u$ is convex (respectively, semiconvex).

Definition 5.1 Given $u : \mathbb{R}^n \to \mathbb{R}$, we say that u is concave if and only if

$$u\big(\lambda p + (1-\lambda) q\big) \geq \lambda u(p) + (1-\lambda) u(q) \quad \forall \lambda \in [0,1] \text{ and } \forall p, q \in \mathbb{R}^n.$$

Recall u is convex if and only if $-u$ is concave.

F. Dragoni (✉)
School of Mathematics, Cardiff University, Cardiff, UK
e-mail: DragoniF@cardiff.ac.uk

M. Chatzakou et al. (eds.), *Direct and Inverse Problems with Applications*, Trends in Mathematics 12, https://doi.org/10.1007/978-3-031-98645-1_5

Example Some very well-known examples are all affine functions, which are both convex and concave on $\mathbb{R}^n$, the modulus $|x|$ is convex, thus $-|x|$ is concave; similarly $u(x) = x^2$ is convex while $u(x) = -x^2$ is concave.

It is also easy to show that whenever $u \in C^2$, then u is concave if and only if $D^2u \le 0$, where this means that the $n \times n$ (symmetric) matrix of second derivatives D^2u is semi-negative definite (i.e. all the eigenvalues are non-positive).

A very important generalization of the notion of concavity is the definition of semiconcavity. We next recall some equivalent notions and refer to [10] for a beautiful overview on this theory. Note also that in this case, u semiconvex if and only if $-u$ semiconcave.

Definition 5.2 (*Semiconcavity*) Given $u : \mathbb{R}^n \to \mathbb{R}$, the following statements are equivalent: $\exists\, C > 0$ such that

1. $p \mapsto u(p) - \frac{C}{2}|p|^2$ is concave.
2. $\forall\, \lambda \in [0, 1],\ p, q \in \mathbb{R}^n,\ \lambda\, u(p) + (1 - \lambda)\, u(q) - u\big(\lambda\, p + (1 - \lambda)\, q\big) \le C|p - q|^2$.
3. $\forall\, p, v \in \mathbb{R}^n, u(p + v) + u(p - v) - 2u(p) \le C|v|^2$ (mid-point semiconcavity).

In this case, the function u is called semiconcave.

If in addition we assume $u \in C^2$, then u is semiconcave if and only if $\exists\, C > 0$ such that $D^2u \le C\ Id_{n\times n}$. Another key property of semiconcave (or semiconvex) functions is that they are locally Lipschitz continuous, which also implies that the gradient is (uniformly) locally bounded.

Example

1. Concave functions are also semiconcave.
2. C^2 functions are both semiconvex and semiconcave in every bounded set.
3. The Euclidean square distance is both semiconcave and semiconvex in $\mathbb{R}^n$ since the Hessian is $2\, Id$, while the distance from the origin is only semiconvex (since it is convex), but it is not semiconcave at the origin.
4. The CC distance from the origin is locally semiconcave outside the origin [9, 20] in ideal Carnot groups (e.g. in the Heisenberg group). □

While convexity and concavity are key geometrical properties for sets and functions, semiconvexity and semiconcavity turn out to be key regularity properties, for example, a function that is both semiconvex and semiconcave is $C^{1,1}$, see [10].

The bounds on their first and second derivatives also have many applications in the theory of PDEs.

In the next session, we characterise concavity and semiconcavity by looking at the second derivatives, also for functions not necessarily C^2, interpreting the sign/bounds for the second derivatives in the viscosity sense.

2 Viscosity Solutions and Semiconcavity

The theory of viscosity solutions has been formally introduced by Crandall and Lions [12, 13] and developed in the following years with the contributions of many other authors. In this section, we recall the main definitions and apply it to a characterization for concave and semiconcave functions.

Definition 5.3 We say that a continuous function $u : \Omega \to \mathbb{R}$ is a

1. viscosity subsolution at some point $x_0 \in \Omega$, if and only if, for any test function $\varphi \in C^2(\Omega)$ such that $u - \varphi$ has a local maximum at x_0, then
$$F(x_0, u(x_0), D\varphi(x_0), D^2\varphi(x_0)) \leq 0.$$
2. viscosity supersolution at some point $x_0 \in \Omega$, if and only if, for any test function $\varphi \in C^2(\Omega)$ such that $u - \varphi$ has a local minimum at x_0, then
$$F(x_0, u(x_0), D\varphi(x_0), D^2\varphi(x_0)) \geq 0.$$
3. viscosity solution at a point $x_0 \in \Omega$ if u is both a viscosity subsolution and a viscosity supersolution.

Example If we consider the eikonal equation $|u'(x)| = 1$ on $(-1, 1)$, with vanishing boundary conditions $u(-1) = 0 = u(1)$, by simply applying Rolle's theorem, one can show that there are no classical solutions, in fact any function $C^1((-1, 1)) \cap C([-1, 1])$ solving the vanishing boundary condition, have at least one point where $u' = 0$. By simply applying the previous definition, it is easy to check that the function $u(x) = -|x| + 1$ is the (only) viscosity solution of the eikonal equation, satisfying the given vanishing boundary conditions. □

Viscosity solutions are a very good notion of generalised solutions for a large class of partial differential equations. In order to show that, we need to verify that they are consistent with classical solutions, which means the following:

1. Whenever $u \in C^2$ is viscosity solution, then u is also classical solution. That is very easy to check since we can take $u = \varphi$ for both the viscosity supersolution and the viscosity subsolution properties, which trivially implies
$$F(x_0, u(x_0), Du(x_0), D^2u(x_0)) = 0, \quad \forall\, x_0 \in \Omega.$$
2. To show that classical solutions are indeed also viscosity solutions is more tricky, and one needs to assume the following additional assumption on the structure of the PDE:
$$F(x, z, p, M) \leq F(x, z, p, N), \quad \forall\, M \geq N, \tag{1}$$
where M and N are symmetric $n \times n$-matrices, therefore $M \geq N$ needs to be interpreted e.g. as all eigenvalues for the matrix $M - N$ are non-negative.

Whenever the PDE satisfies the previous property of monotonicity w.r.t. the second derivatives, the PDE is called *degenerate elliptic*. Assuming this additional property, we can show that classical solutions are also viscosity solutions, in fact, given $\varphi \in C^2(\Omega)$ such that $u - \varphi \in C^2(\Omega)$ has a local maximum at x_0, then

$$\begin{cases} D(u-\varphi)(x_0) = 0, \\ D^2(u-\varphi)(x_0) \leq 0. \end{cases}$$

Hence u classical solution, together with property (1), give

$$F(x_0, u(x_0), Du(x_0), D^2u(x_0)) = 0 \quad \Rightarrow F(x_0, u(x_0), D\varphi(x_0), D^2\varphi(x_0)) \leq 0,$$

i.e. the viscosity subsolution property. Similarly, one can check the viscosity supersolution property.

Example

1. First-order PDEs (for example, the eikonal equation or, in general, Hamilton-Jacobi equations) are always degenerate elliptic.
2. Linear elliptic and parabolic equations (e.g. Poisson equations or heat equations) are degenerate elliptic.
3. Hyperbolic equations (such as the wave equation) are not degenerate elliptic.

Viscosity solutions usually have very good existence, uniqueness and stability properties. For detailed theorems and suitable assumptions, we refer to [3, 6, 8, 11, 16, 19, 21] which give a complete overview of the theory from different point of views.

We can now use the theory of viscosity solutions to characterise concave and semiconcave functions.

Theorem 5.1 ([1]) *Let u lower semicontinuous on $\Omega \subset \mathbb{R}^n$ open, then u is concave, if and only if, $-D^2u(x) \geq 0$, in the viscosity sense, in Ω. This means that $D^2\phi(x_0) \geq 0$ for all smooth $\phi \in C^2(\Omega)$ such that $u - \varphi$ has a local minimum at some point $x_0 \in \Omega$ (viscosity supersolution property).*

The previous theorem can be applied to semiconcave functions: this means that, given u lower semicontinuous on $\Omega \subset \mathbb{R}^n$ open, then u is semiconcave, if and only if,

$$-D^2u(x) \geq -C\, Id_{n\times n}, \quad \text{in the viscosity sense, in } \Omega.$$

3 Carnot Groups

In this session, we are going to give a brief overview of Carnot groups, recalling in particular the main definitions and properties. In particular, we focus on the gauge (also called homogeneous) distance and the Carnot-Carathéodory (shortly CC) distance.

Definition 5.4 A Carnot group $(\mathbb{G}, \cdot)$ is a simply connected nilpotent Lie group, whose Lie algebra g has a finite stratification, i.e.

$$\mathrm{g} = \bigoplus_{j=1}^{r} \mathrm{g}_j, \quad \text{i.e.} \quad \begin{cases} [\mathrm{g}_1, \mathrm{g}_j] = \mathrm{g}_{j+1}, \ \forall\, j = 1, \ldots, r-1, \\ [\mathrm{g}_1, \mathrm{g}_r] = \{0\}. \end{cases}$$

If $\mathrm{g}_r \neq \{0\}$, then r is the step of the Carnot group $\mathbb{G}$.

Example The n-dimensional Heisenberg group $\mathbb{H}^n$ is a step 2 Carnot group defined on $\mathbb{R}^{2n+1}$ by the following non-abelian operation:

$$p \cdot q = \left(\overline{x}_1 + \overline{y}_1, \overline{x}_2 + \overline{y}_2, x_3 + y_3 + \frac{\langle \overline{x}_1, \overline{y}_2 \rangle - \langle \overline{x}_2, \overline{y}_1 \rangle}{2} \right),$$

where $p = (\overline{x}_1, \overline{x}_2, x_3), q = (\overline{y}_1, \overline{y}_2, y_3) \in \mathbb{R}^n \times \mathbb{R}^n \times \mathbb{R}$, when $+$ denotes the Euclidean sum and $\langle \, , \, \rangle$ the standard inner product in $\mathbb{R}^n$. □

For all $\lambda > 0$, we consider the automorphism $\Delta_\lambda : \mathrm{g} \to \mathrm{g}$ defined by $\Delta_\lambda(X) = \lambda^i X$, for all $X \in \mathrm{g}_i$ and $i = 1, \ldots, r$. By using the finite stratification of g, we can deduce:

$$\Delta_\lambda(X) = \lambda X_1 + \cdots + \lambda^r X_r, \quad \text{for } X = X_1 + \cdots + X_r \text{ and } X_i \in \mathrm{g}_i.$$

On a Carnot group $\mathbb{G}$, the exponential map is a diffeomorphism from g to $\mathbb{G}$ (see [7] for a definition and properties). Therefore, we can define the dilations on the group $\mathbb{G}$ as the smooth map $\delta_\lambda : \mathbb{G} \to \mathbb{G}$ given by $\delta_\lambda = exp \circ \Delta_\lambda \circ exp^{-1}$, where by exp we indicate the exponential map. Note that the dilations give a family of anisotropic scalings whenever $r \geq 2$.

Using again the stratification of g and the exponential map, we can introduce a gauge (or homogeneous) norm as

$$\|x\|_G := \left(\sum_{i=1}^{r} |\overline{x}_i|^{\frac{2r!}{i}} \right)^{\frac{1}{2r!}}, \quad x = (\overline{x}_1, \ldots, \overline{x}_r) \in \mathbb{G}.$$

The gauge norm rescales with the dilations: $\|\delta_\lambda(x)\|_G = \lambda \, \|x\|_G$. Note that $\|x\|_G^{2r!}$ is a polynomial, so it is always smooth (in the standard sense). As already remarked smooth functions are trivially both semiconvex and semiconcave in every bounded set. Given $p, q \in \mathbb{G}$, the gauge distance is $d_G(p, q) = \left\| q^{-1} \cdot p \right\|_G$.

To introduce the CC distance on Carnot groups, we need to consider the sub-Riemannian manifold structure of canonical left-invariant vector fields defined on Carnot groups. We refer to [7, 25] for details. Here, we just recall that we can identify the horizontal distribution $\mathcal{D} \subset T\mathbb{G}$ generated by g_1 by using a suitable family of left-invariant vector fields $X_1, \ldots, X_m$ where $m = \dim \mathrm{g}_1$.

The key notion necessary to introduce the CC distance is the one of admissible (horizontal) curves: a horizontal curve γ is every absolutely continuous curve defined on the group, such that $\dot{\gamma}(t) \in \mathcal{D}_{\gamma(t)}$, a.e. t, that means $\dot{\gamma}(t) = \alpha_1(t)X_1(\gamma(t)) + \cdots + \alpha_m(t)X_m(\gamma(t))$, for $\alpha_1, \ldots, \alpha_m$ suitable measurable functions and $X_1, \ldots, X_m$ generating $\mathcal{D}$. On $\mathcal{D}$ we can define a Riemannian metric $g(\cdot, \cdot)$, so for horizontal curves, the length can be introduced as $l(\gamma) = \int_0^1 \sqrt{g(\dot{\gamma}(t), \dot{\gamma}(t))}\, dt$.

Definition 5.5 Given $p, q \in \mathbb{G}$, the Carnot-Carathéodory (CC) distance is

$$d(p,q) = \inf \left\{ l(\gamma) \mid \gamma : [0,1] \to \mathbb{G}, \text{ horizontal } \gamma(0) = p, \gamma(1) = q \right\}.$$

4 h-Semiconcavity and the Square CC Distance

In this section we recall how to adapt the previous definitions and properties of concave and semiconcave functions to Carnot groups, and in particular the viscosity characterization by the intrinsic (horizontal) second derivatives. The horizontal convexity (shortly h-convexity) was first introduced in [14, 22, 23], and later generalised to the case of general Hörmander vector fields in [4, 5]. We also refer to [17] for an overview of the subject. As done for the Euclidean counterpart, in this section, we write all definitions and properties directly for the h-concave, respectively h-semiconcave, case.

Definition 5.6 Given $\Omega \subset \mathbb{G}$ open and u lower semicontinuous, u is h-concave if and only if

$$u(p \cdot h) + u(p \cdot h^{-1}) - 2u(p) \leq 0,$$

$\forall\, p \in \Omega, h \in \mathcal{H}_0 = \mathcal{D}_0$ such that $[p \cdot h^{-1}, p \cdot h] := \{p \cdot \tau h \mid \tau \in [-1, 1]\} \subset \Omega$.

In step 2 Carnot groups, (Euclidean) concavity implies h-concavity.

Recall that, given $\mathcal{X} = \{X_1, \ldots, X_m\}$ a generic family of (left-invariant) vector fields, the intrinsic gradient is given by

$$D_{\mathcal{X}} u = (X_1 u, \ldots, X_m u)^t = \sigma D u,$$

where $\sigma = [X_1, \ldots, X_m]^t$ is the $m \times n$ matrix associated to the family $\mathcal{X}$.

The intrinsic Hessian is given by the $m \times m$-matrix

$$D^2_{\mathcal{X}} u = \big(X_i(X_j u)\big)_{i,j=1}^m.$$

Since in general $D^2_{\mathcal{X}} u$ may not be symmetric, we consider the symmetrized matrix

$$\left(D^2_{\mathcal{X}} u\right)^* = \left(\frac{X_i(X_j u) + X_j(X_i u)}{2}\right)_{i,j=1}^m$$

$$= \sigma\, D^2 u\, \sigma^t + \left(\frac{\nabla_{X_j} X_i + \nabla_{X_i} X_j}{2} \cdot Du \right)_{i,j=1}^{m}.$$

Using the horizontal derivatives, we can generalise the viscosity characterization for concave functions to the case of h-concave functions. The following theorem has been proved in Carnot groups in [22, 23, 26] and later generalized to the case of Hörmander vector fields in [4].

Theorem 5.2 *Let $u \in C(\mathbb{R}^n)$, then u is h-concave, if and only if, $-\big(D^2_{\mathcal{X}} u\big)^* \geq 0$, in the viscosity sense, in $\mathbb{R}^n$.*

The previous definition and result can be generalised to the smiconcave case.

Definition 5.7 $u : \mathbb{G} \to \mathbb{R}$ is h-semiconcave if $\exists\, C > 0$ such that $\forall p \in \mathbb{G}$ and $\forall h \in \mathcal{H}_0$, $u(p \cdot h) + u(p \cdot h^{-1}) - 2u(p) \leq C|h|^2$. Note that $h \in \mathcal{H}_0 \Leftrightarrow h = (h_1, \dots h_m, 0, \dots, 0)$ with $m = dim(\mathfrak{g}_1)$, that implies $\|h\|_G = |h|$ (Euclidean norm).

By using the previous theorem, it is trivial to show that, given $u : \mathbb{G} \to \mathbb{R}$ lower semicontinuous, u is h-semiconcave if and only if $-\big(D^2_{\mathcal{X}} u\big)^* \geq -C\, Id_{m \times m}$ in the viscosity sense.

By using the viscosity characterization, in [18], we could prove that the square CC distance from the origin is h-semiconcave in the whole space. First, we want to recall that in [9, 20], it was proved that in the ideal group, the CC distance from the origin is locally Euclidean semiconcave outside the origin. Since ideal groups are a subfamily of step 2 Carnot groups, this result implies local h-semiconcavity outside the origin. While the dilations can be used to extend the horizontal (not the Euclidean) property at infinity, the result can be proved to be false at the origin (see [18] for more details). Nevertheless, the result can be used to prove the following theorem in the ideal group. For non ideal step 2 groups, one can instead use the recently developed notion of C-nearly horizontal semiconcavity and the related results in [2].

Theorem 5.3 ([18] *In all step 2 groups $\mathbb{G}$, $d^2(p, 0)$ is h-semiconcave in $\mathbb{G}$.*

Note that the result is optimal in the sense that we can show that in the Engel group (simplest example step 3 Carnot group), no powers of the CC distance are h-semiconcave near the origin. Moreover, different from the Euclidean case, one can show that in the Heisenberg group, the square CC distance from the origin is not h-semiconvex.

In [18], the previous theorem, together with the stability of viscosity supersolution w.r.t. infimum, have also been used to prove h-semiconcavity properties for the viscosity solutions of a large class of degenerate Hamilton-Jacobi equations, by using the metric Hopf-Lax formula developed in [15, 24].

References

1. Alvarez, O., Lasry, J.-M., Lions, P.-L.: Convex viscosity solutions and state constraints. Journal de Mathématiques Pures et Appliquées **76**(3), 265–288 (1997)
2. Badreddine, Z., Rifford, L.: Measure contraction properties for two-step analytic sub-Riemannian structures and Lipschitz Carnot groups. Ann. Inst. Fourier **70**(6), 2303–2330 (2020)
3. Bardi, M., Capuzzo Dolcetta, I.: Optimal Control and Viscosity Solutions of Hamilton-Jacobi-Bellman Equations. Birkhäuser, Basel (1997)
4. Bardi, M., Dragoni, F.: Convexity and semiconvexity along vector fields. Calculus of Variations and Partial Differential Equations **42**(3–4), 405–427 (2011)
5. Bardi, M., Dragoni, F.: Subdifferential and properties of convex functions with respect to vector fields. J. Convex Anal. **21**(3), 785–810 (2014)
6. Barles, G.: Solutions de Viscosité des Equations de Hamilton-Jacobi. Springer, Berlin (1994)
7. Bonfiglioli, A., Lanconelli, E., Uguzzoni, F.: Stratified Lie Groups and Potential Theory for their Sub-Laplacians. Springer, Berlin (2007)
8. Caffarelli, L., Cabré, X.: Fully Nonlinear Elliptic Equations. American Mathematical Society, Providence (1995)
9. Cannarsa P., Rifford, L.: Semiconcavity results for optimal control problems admitting no singular minimizing controls. Ann. Inst. H. Poincaré C Anal. Non Linéaire, **25**(4), 773–802 (2008)
10. Cannarsa, P., Sinestrari, C.: Semiconcave Functions, Hamilton-Jacobi Equations, and Optimal Control. Birkhäuser, Basel (2004)
11. Crandall, M.G., Ishii, H., Lions, P.-L.: User's guide to viscosity solutions of second order partial differential equations. Bull. AMS **27**(1), 1–67 (1992)
12. Crandall, M.G., Lions, P.L.: Condition d'unicité pour les solutions generalisées des équations de Hamilton-Jacobi du premier ordre. Acad. Sci. Paris Sér. I Math **292**, 183–186 (1981)
13. Crandall, M.G., Lions, P.L.: Viscosity solutions of Hamiltonian-Jacobi equations. Trans. Am. Math. Soc. **277**, 1–42 (1983)
14. Danielli, D., Garofalo, N., Nhieu, D.-M.: Notions of convexity in Carnot groups. Comm. Anal. Geom. **11**(2), 263–341 (2003)
15. Dragoni, F.: Metric Hopf-Lax formula with semicontinuous data. Discrete Continuous Dyn. Syst. **17**(4), 713–729 (2007)
16. Dragoni, F.: Introduction to viscosity solutions for nonlinear PDEs (2012). Available online
17. Dragoni, F., Filali, D.: Starshaped and convex sets in Carnot groups and in the geometries of vector fields. J. Convex Anal. **4**(24), 1349–1372 (2019)
18. Dragoni, F., Liu, Q., Zhang Y.: Horizontal semiconcavity for the square of Carnot-Carathéodory distance on step 2 Carnot groups and applications to Hamilton-Jacobi equations. Preprint (2024)
19. Evans, L.C.: Partial Differential Equations. American Mathematical Society, Providence (2010)
20. Figalli, A., Rifford, L.: Mass transportation on sub-Riemannian manifolds. Geom. Funct. Anal. **20**(1), 124–159 (2010)
21. Giga, Y.: Surface Evolution Equations A Level Set Approach. Birkhäuser, Basel (2002)
22. Juutinen, P., Lu, G., Manfredi, J., Stroffolini, B.: Convex functions on Carnot groups. Rev. Mat. Iberoamericana **23**(1), 191–200 (2007)
23. Lu, G., Manfredi, J., Stroffolini, B.: Convex functions in Heisenberg. Group. Cal. Var. **19**, 1–22 (2004)
24. Manfredi, J., Stroffolini, B.: A version of the Hopf-Lax formula in the Heisenberg group. Commun. Part. Diff. Eqn. **27**, 1139–1159 (2002)
25. Montgomery, R.: A Tour of Subriemannian Geometries. Their Geodesics and Application. American Mathematical Society, Providence (2002)
26. Wang, C.: Viscosity convex functions on Carnot Groups. Proc. Amer. Math. Soc. **133**, 1247–1253 (2005)

Direct and Inverse Nonstationary Scattering Problems for Dirac-Type System

Mansur I. Ismailov

Abstract This paper examines direct and inverse nonstationary scattering problems for the Dirac-type system on both the whole axis and the semi-axis. It defines the scattering operators for the nonstationary Dirac-type system. Using the Gelfand-Levitan-Marchenko type equations, we prove the unique reconstruction of the potential from the scattering operator on the whole axis. Additionally, we establish the uniqueness of the inverse scattering problem for the Dirac-type system on the semi-axis by the relationship between the scattering operators on the whole and semi-axis.

Keywords Nonstationary scattering problem · Dirac-type system · Inverse problem

MSC2020 Primary 35R30 · Secondary 35L50 · 35P25 · 37K15

1 Introduction

In contrast to the traditional underlying model for wave propagation, it is observed in [2] that for problems involving wave propagation in non-stationary media, the coefficients of the relevant hyperbolic partial differential equations vary not only in space but also in time. While there is a substantial amount of literature on the inverse scattering problem (ISP) for determining coefficients that depend solely on space, the literature on ISP for models involving non-stationary media remains relatively limited.

Let us consider the system of first-order hyperbolic equations

$$\partial_t \psi - \sigma \partial_x \psi = [\sigma, Q]\, \psi, \tag{1}$$

M. I. Ismailov (✉)
Department of Mathematics, Gebze Institute of Technology, Gebze-Kocaeli, Turkey
e-mail: mismailov@gtu.edu.tr

Center for Mathematics and its Applications, Khazar University, Baku, Azerbaijan

M. Chatzakou et al. (eds.), *Direct and Inverse Problems with Applications*,
Trends in Mathematics 12, https://doi.org/10.1007/978-3-031-98645-1_6

where $[\sigma, Q] = \sigma Q - Q\sigma$ is the commutator, $\sigma = diag\{\xi_1, \xi_2, \ldots, \xi_n\}$ is a constant diagonal matrix with nonzero real diagonal elements, $[\sigma, Q(x,t)]$ is an $n \times n$ matrix with measurable, complex-valued, square-summable entries. We refer to this matrix function as the potential. The ISP for time-dependent potentials of (1) was started by Nizhnik [13].

Because the potential is time-dependent, the system (1) describes interacting waves with velocities $\xi_1, \xi_2, \ldots \xi_n$ propagating in a non-stationary medium. Moreover, motion from right to left corresponds to positive velocity values, while motion from left to right corresponds to negative velocity values. In the case where $[\sigma, Q] = 0$, the system (1) describes the propagation of n free waves with velocities $\xi_1, \xi_2, \ldots \xi_n$.

In the case of distinct velocities $\xi_i \neq \xi_j$ for $i \neq j$, the ISP for the system (1) on the whole plane was studied in [16] using a Gelfand-Levitan-Marchenko (GLM) type linear integral equation, and in [19] using a Riemann-Hilbert problem. For the case where the velocities are equal, the problem for the system (1) was studied in [7] using a GLM type linear integral equation.

For the system (1) on the half-axis, some components of the velocities must have the opposite sign to the others for the scattering process to occur. Let the velocities satisfy the following condition:

$$\xi_1 \geq \cdots \geq \xi_p > 0 > \xi_{p+1} \geq \cdots \geq \xi_n,$$

where $1 \leq p < n$. This implies that the system (1 describes wave propagation in a nonstationary medium with p incident waves and $n - p$ scattered waves. The scattering problems on the half-plane pose fundamental challenges in determining the sufficient number of such problems (on the semi-axis for the same system of equations) required to guarantee a unique solution for the ISP under consideration. The problems for which uniqueness results are established correspond to the following cases:

(1) When $n = 2m$ and $p = m$, the system (1) implies an equal number (m) of incident and scattered waves. In this case, the ISP was studied in [8] for equal velocities of incident and scattered waves using a single scattering problem, and partially in [9] for different velocities of incident and scattered waves using a two-scattering problem.
(2) When $n = m + 1$ and $p = m$, the system (1) implies m incident waves and a single scattered wave. In this case, the ISP was studied in [10] for equal velocities of the incident waves, and in [11, 17] for different velocities of the incident waves, using a m-scattering problem.

The case where the potential is solely space-dependent was studied in [3] for the ISP of the two-velocity system (the case $\xi_1 > \xi_2 > 0 > -\xi_2 > -\xi_3$) on the half-axis, considering a symmetric potential localized on a finite interval. Additionally, the numerical aspects of the ISP for the system (1), in the general case of incident and scattered waves with different velocities, were examined in [1] for a potential that is solely space-dependent. In the case $n = 2$, $p = 1$, the classical nonstationary Dirac system has a rich literature on the inverse spectral problem (ISP) on both the whole and half-axis [4, 5, 13, 18].

In this paper, we consider the direct and inverse scattering problems for system (1) on both the whole axis and the half-axis in the case $\xi_1 = \xi_2 = \cdots = \xi_m > 0 > \xi_{m+1}$. On the half-axis, the system describes m incident waves (with equal velocities) and a single scattered wave. We study this system under a more general boundary condition. We recall the uniqueness results of the ISP for a system (1) on the whole axis [7], and we determine the uniqueness of the ISP for a system (1) on the semi-axis by the relation between the scattering operators on the whole and half-axis.

Notation: In this article, the following Volterra type integral operators will be used:

$$\mathbf{A}_{\pm}(x)\, h(x,t) = \mp \int_{t}^{\mp\infty} \mathbf{A}_{\pm}(x,t,\tau)\, h(x,\tau)\, d\tau .$$

We will denote by

$$\mathbf{A}_{-} = \mathbf{A}_{-}(0) \text{ and } \mathbf{A}_{+} = \mathbf{A}_{+}(0).$$

The following facts are about the right and left factorizations of the Hilbert-Schmidt integral operators of the second kind in $\mathbf{L}_2(\mathbb{R}, \mathbb{C}^n)$. These results immediately follow from the Theorems 4.2.2 and 4.2.3 in [14] and will be used below.

Let $\mathbf{F}$ is a Hilbert-Schmidt integral operator and $\mathbf{S} = \mathbf{I} + \mathbf{F}$, where $\mathbf{I}$ is the identity operator in $\mathbf{L}_2(\mathbb{R}, \mathbb{C}^n)$. If the operator $\mathbf{S}$ admits right (or left) factorization as $\mathbf{S} = (\mathbf{I} + \mathbf{A}_{+})(\mathbf{I} + \mathbf{A}_{-})$ (or $\mathbf{S} = (\mathbf{I} + \mathbf{A}_{-})(\mathbf{I} + \mathbf{A}_{+})$), then the operators $\mathbf{A}_{+}$ and $\mathbf{A}_{-}$ uniquely determine $\mathbf{S}$ with respect to $\mathbf{F}$. This determination corresponds to a Gelfand-Levitan-Marchenko-type linear integral equation, which is the main equation in inverse scattering theory.

The interested readers can also refer to [6] for necessary facts on abstract Volterra type operators and on factorizations of the second kind Fredholm integral operators.

2 Direct and Inverse Scattering Problem on the Whole Plane

Let $\sigma = \operatorname{diag}\{1, \ldots, 1, -1\}$ in the system (1).

2.1 *Scattering Problem on the Whole Plane*

In this chapter, the system (1) will be considered on the plane

$$\mathbb{R}^2 = \{(x,t) : -\infty < x, t < +\infty\}.$$

If $[\sigma, Q] = 0$, then we will call the system (1) a nonperturbated system. It is easy to verify that the locally integrable vector-function ψ satisfies the nonperturbed system (1) in a generalized sense if and only if it is given by

$$\psi(x,t) = F_x h(t)$$

where $h = \text{col}\ [a_1, \ldots, a_m, b]$ are arbitrary locally integrable vector-functions and F_x is the sifted operator: $F_x h(t) = \text{col}\ [a_1(t+x), \ldots, a_m(t+x), b(t-x)]$.

We will consider generalized solutions

$$\psi(x,t) = \text{col}[\varphi_1(x,t), \ldots, \varphi_m(x,t), \varphi_{m+1}(x,t)]$$

of the system (1), which are measurable in x and t. Here, these functions $\varphi_1(x,t), \ldots, \varphi_m(x,t)$ and $\varphi_{m+1}(x,t)$ belong to space $\mathbf{L}_2(\mathbb{R}, \mathbb{C})$ with respect to variable t and their $\mathbf{L}_2$ -norms in time belong to $\mathbf{L}_\infty(\mathbb{R}, \mathbb{C})$. We call such solutions admissible on the whole axis.

Theorem 6.1 *Let the coefficients of the system (1) be measurable, complex-valued and square-summable functions. The following statements hold:*

(a) for an arbitrary vector-function $h_- \in \mathbf{L}_2\left(\mathbb{R}, \mathbb{C}^{m+1}\right)$ there exists a unique admissible solution ψ of the system (1), satisfying the asymptotic

$$\psi(x,t) = F_x h_-(t) + o(1) \text{ as } t \to -\infty; \tag{2}$$

(b) for an arbitrary admissible solution ψ of the system (2) there exists a unique vector-function $h_+ \in \mathbf{L}_2\left(\mathbb{R}, \mathbb{C}^{m+1}\right)$ satisfying the asymptotic

$$\psi(x,t) = F_x h_+(t) + o(1) \text{ as } t \to +\infty. \tag{3}$$

The problem (1), (2) is equivalent to following integral equation:

$$\psi(x,t) = F_x h_-(t) + \int_{-\infty}^{t} F_{t-s}([\sigma, Q]\psi)(x,s)\,ds. \tag{4}$$

This integral equation has a unique admissible solution since it is Volterra integral equation in the variable t, and its kernel is square summable.

If ψ is an admissible solution of the system (1), then direct checking shows that the quantity

$$\varphi(x,t) = \psi(x,t) + \int_{t}^{+\infty} F_{t-s}([\sigma, Q]\psi)(x,s)\,ds$$

satisfies the unperturbed system (1). Therefore, there exists a vector function $h_+ \in \mathbf{L}_2\left(\mathbb{R}, \mathbb{C}^{m+1}\right)$ such that $\varphi(x,t) = F_x h_+(t)$ i.e.

$$\psi(x,t) = F_x h_+(t) - \int_t^{+\infty} F_{t-s}\left([\sigma, Q]\,\psi\right)(x,s)\,ds. \tag{5}$$

The asymptotic relation (3) easily follows from (5).

The vector-function $h_- \in \mathbf{L}_2\left(\mathbb{R}, \mathbb{C}^{m+1}\right)$ defines the profile of the incident waves. The scattering problem consists in finding the solution to the system (1) when the incident waves are given. The first part of Theorem 6.1 gives the solution of the scattering problem for the system (1). This solution determines a vector-function $h_+ \in \mathbf{L}_2\left(\mathbb{R}, \mathbb{C}^{m+1}\right)$ according to the second part of the Theorem 6.1. The vector-function $h_+ \in \mathbf{L}_2\left(\mathbb{R}, \mathbb{C}^{m+1}\right)$ defines the profile of the scattered waves. It can be obtained from the equalities (4) and (5) and it is as follows

$$F_x h_+(t) = F_x h_-(t) + \int_{-\infty}^{+\infty} F_{t-s}\left([\sigma, Q]\,\psi\right)(x,s)\,ds \tag{6}$$

The operator $\mathbf{S}$ which transforms the incident waves h_- into the scattered waves h_+ is called the scattering operator for the system (1) on the plane:

$$h_+(t) = \mathbf{S}h_-(t). \tag{7}$$

The operator $\mathbf{S}$ is a $(m+1)\times(m+1)$ matrix integral operator in the space $\mathbf{L}_2\left(\mathbb{R}, \mathbb{C}^{m+1}\right)$.

2.2 *Transformation Operators*

In solving the inverse scattering problems, the Volterra-type integral representation of the solution plays an important role. Such representation can be taken from transforming operators as $t \to -\infty$ and $t \to +\infty$.

Finding the admissible solution to the system (1) with the given asymptotic $F_x h_-(t)$ as $t \to -\infty$ is equivalent to the solvability of the system of integral equation (4). Alternatively, the system (1) with the given asymptotic $F_x h_+(t)$ as $t \to +\infty$ is equivalent to the solvability of the integral equation (5).

We will consider the integral equations (4) and (5) with arbitrary given free term.

Lemma 6.1 *For any $h_\pm \in \mathbf{L}_2\left(\mathbb{R}, \mathbb{C}^{m+1}\right)$ there exists a unique admissible solution of the systems (4) and (5), and these solutions admit the representations*

$$\psi(x,t) = \left[\mathbf{I} + \mathbf{A}_\pm(x)\right] F_x h_\mp(t), \tag{8}$$

where $\mathbf{I}$ *is the identity operator, the matrix elements* $A_{ij}^{\pm}(x,t,\tau)$, $i,j=1,2,3$, *of the kernel of the integral operator* $\mathbf{A}_{\pm}(x)$ *are determined uniquely by the coefficients of the system (2), and for the fixed* x *these kernels are the Hilbert-Schmidt kernels. In addition, these kernels are connected with the potential by the following equality*

$$[\sigma, Q](x,t) = \pm\left[\sigma, A^{\pm}\right](x,t,t). \tag{9}$$

The existence and uniqueness of the admissible solution of the systems (4) and (5) are proved in Theorem 6.1. If the solution of (4) can be represented as in the form of (5) for each $h_{\pm} \in \mathbf{L}_2$, then substituting (8) in (4), we obtain the system of integral equations for the kernels $A_{ij}^{\pm}(x,t,\tau)$ and the reconstruction formula (9).

2.3 Inverse Scattering Problem

From the Lemma 6.1, we obtain that

$$\left[\mathbf{I}+\mathbf{A}_{-}(x)\right] F_x h_{+}(t) = \left[\mathbf{I}+\mathbf{A}_{+}(x)\right] F_x h_{-}(t). \tag{10}$$

If we consider that $\left[\mathbf{I}+\mathbf{A}_{-}(x)\right]^{-1}$ exists and $F_x^{-1} = F_{-x}$ then from (10) and the definition of the scattering operator (7), we take the following representation for the scattering operator.

Theorem 6.2 *Let the operator* $\mathbf{S}$ *be the scattering operator of the scattering problem for the system (1) on the whole axis. Then for every* x, *the operator* $F_x\mathbf{S}F_{-x}$ *admits the following factorizations*

$$F_x\mathbf{S}F_{-x} = \left[\mathbf{I}+\mathbf{A}_{-}(x)\right]^{-1}\left[\mathbf{I}+\mathbf{A}_{+}(x)\right].$$

In particular, when $x=0$, *the operator* $\mathbf{S}$ *admits a left-sided factorization.*

By virtue of the uniqueness of factorization in terms of Volterra factors, we derive the following result regarding the inverse scattering problem for system (1) on the whole axis.

Theorem 6.3 *Let the operator* $\mathbf{S}$ *be the scattering operator for the system (1) on the entire axis, with a measurable, complex-valued, and square-summable potential* $[\sigma, Q]$. *Then the potential* $[\sigma, Q]$ *is uniquely determined by the given scattering operator* $\mathbf{S}$.

3 Direct and Inverse scattering Problem on Half-Plane

Let us denote by

$$\psi = \text{col}\left[\varphi_1, \ldots, \varphi_m, \varphi_{m+1}\right].$$

Then, we can rewrite the system (2) in the following form

$$\begin{aligned} \frac{\partial \varphi_i}{\partial t} - \frac{\partial \varphi_i}{\partial x} &= p_i(x,t)\,\varphi_{m+1}, \quad i = 1, 2, \ldots, m, \\ \frac{\partial \varphi_{m+1}}{\partial t} + \frac{\partial \varphi_{m+1}}{\partial x} &= p_{m+1}(x,t)\,\varphi_1 + \cdots + p_{2m}(x,t)\,\varphi_m. \end{aligned} \tag{11}$$

3.1 Scattering Problem on Half-Plane

The system (11) will be considered on the half-plane

$$\mathbb{R}^2_+ = \{(x,t) : x \geq 0, -\infty < t < +\infty\}.$$

The nonstationary scattering problem for the system (11) on the half-plane consists of finding the solution of the system (11) with respect to the given asymptotic relations

$$\varphi_i(x,t) = a_i(x+t) + o(1), \quad x \to +\infty\,, \quad i = 1, 2, \ldots, m, \tag{12}$$

and the boundary condition at $x = 0$, where the functions $a_1, a_2, \ldots, a_m$ define the profile of the incident waves.

Let us consider m scattering problems for the system (11) on the half-axis: The k-th problem consists of finding the solution

$$\psi^k(x,t) = \text{col}[\varphi_1^k(x,t), \ldots, \varphi_m^k(x,t), \varphi_{m+1}^k(x,t)]$$

of the system (11), with respect to boundary condition

$$\varphi_{m+1}^k(0,t) = h_{k1}\varphi_1^k(0,t) + \cdots + h_{km}\varphi_m^k(0,t) \tag{13}$$

and for given incident waves $a_1, \ldots, a_m \in \mathbf{L}_2(\mathbb{R}, \mathbb{C})$ which define the asymptotics of $\varphi_1^k(x,t), \ldots, \varphi_m^k(x,t)$ as $x \to +\infty$ in the form (12). Here $|h_{k1}| + \cdots + |h_{km}| \neq 0$ for each $k = 1, 2, \ldots, m$.

We will consider generalized solutions of the system (11), which are measurable in x and t. Here, these functions belong to the space $\mathbf{L}_2(\mathbb{R}, \mathbb{C}^{m+1})$ with respect to variable t and their $\mathbf{L}_2$-norms belong to $\mathbf{L}_\infty(\mathbb{R}_+, \mathbb{C}^{m+1})$, i.e. the vector-valued functions $\psi = \text{col}[\varphi_1, \ldots, \varphi_m, \varphi_{m+1}]$ are in the space of vector-valued functions with norm

$$\| \psi \| = \max_{i=1,2,\ldots,m+1} \sup_{x \geq 0} \| \varphi_i (x, \cdot) \|_{\mathbf{L}_2(\mathbb{R})}.$$

We call such solutions admissible.

Theorem 6.4 *If the coefficients of the system (11) are measurable, complex-valued, and square-summable functions, then for given arbitrary incident waves* $a_1, \ldots, a_m \in \mathbf{L}_2 (\mathbb{R}, \mathbb{C})$, *there exists a unique admissible solution of the kth scattering problem, and the last component of the solutions satisfy the asymptotic relations*

$$\varphi_{m+1}^k (x, t) = b_k(t - x) + o(1), \quad x \to +\infty, \quad k = 1, 2, \ldots, m, \tag{14}$$

where all $b_k \in \mathbf{L}_2 (\mathbb{R}, \mathbb{C})$ *define the profile of the scattered wave.*

The kth ($k = 1, 2, \ldots, m$) scattering problem is equivalent to following system of integral equations:

$$\varphi_i (x, t) = a_i(t + x) + \int_x^{+\infty} \left[p_i \varphi_{m+1} \right] (s, t + x - s) \, ds, \quad i = 1, 2, \ldots, m,$$

$$\varphi_{m+1} (x, t) = b_k(t - x) - \int_x^{+\infty} \left[p_{m+1} \varphi_1 + \cdots + p_{2m} \varphi_m \right] (s, t - x + s) \, ds,$$

where

$$\begin{aligned} b_k(t) &= h_{k1} a_1(t) + \cdots + h_{km} a_m(t) \\ &+ \int_0^{+\infty} \Big\{ \left[h_{k1} p_1 \varphi_{m+1} + \cdots + h_{km} p_m \varphi_{m+1} \right] (s, t - s) \\ &+ \left[p_{m+1} \varphi_1 + \cdots + p_{2m} \varphi_m \right] (s, t + s) \Big\} ds, \quad k = 1, 2, \ldots, m. \end{aligned}$$

This integral equation has a unique admissible solution since it is a Volterra type in the variable x, and its kernel is square summable.

According to Theorem 6.4, for an arbitrary $a = \operatorname{col} [a_1, \ldots, a_m] \in \mathbf{L}_2 (\mathbb{R}, \mathbb{C}^m)$, there exist a unique admissible solution of the kth ($k = 1, 2$) scattering problem. Moreover, this solution satisfies the asymptotic relation (14).

Thus for every incident wave vector $a = \operatorname{col} [a_1, \ldots, a_m] \in \mathbf{L}_2$ there are two bounded solutions of the system (11) and these solutions define the profile $b = \operatorname{col} [b_1, \ldots, b_m] \in \mathbf{L}_2$ of the two scattered waves by (14). We can define the operator $\mathbf{S}_H = \left[\mathbf{s}_{ij} \right]_{i,j=1}^m$ translating a to b:

$$b = \mathbf{S}_H a. \tag{15}$$

We will consider these m scattering problems together under the assumption that $\det H \neq 0$, where $H = \left[h_{ij}\right]_{i,j=1}^{m}$. The operator $\mathbf{S}_H$ will be called the scattering operator for the system (11) on the half-plane. This operator is a bounded linear operator on the space $\mathbf{L}_2\left(\mathbb{R}, \mathbb{C}^m\right)$.

3.2 Properties of Scattering Operator on the Half-Plane

By considering the boundary conditions (13) and utilizing the transformation operators (8), we derive the following relationship between the scattering and incident waves:

$$b_k\left(t\right) = \left(\mathbf{I} + \mathbf{M}_{k+}\right)^{-1}\left[\left(h_{k1}\mathbf{I} + \mathbf{M}_{k1-}\right)a_1\left(t\right) + \cdots + \left(h_{km}\mathbf{I} + \mathbf{M}_{km-}\right)a_m\left(t\right)\right],$$
$$k = 1, 2, \ldots, m. \tag{16}$$

where

$$\mathbf{M}_{k+} = \mathbf{A}_{m+1,m+1+} - \sum_{i=1}^{m} h_{ki}\mathbf{A}_{(i,m+1)+}, \quad \mathbf{M}_{kj-} = \mathbf{A}_{m+1,j-} - \sum_{i=1}^{m} h_{ki}\mathbf{A}_{ij-}.$$

By comparing formulas (16) and the definition (15) of the scattering operator, we take the following Volterra structure of the elements of the matrix scattering operator for the system (11) on the half-plane

$$\mathbf{s}_{kj} = \left(\mathbf{I} + \mathbf{M}_{k+}\right)^{-1}\left(h_{kj}\mathbf{I} + \mathbf{M}_{kj-}\right), k, j = 1, 2, \ldots, m.$$

These formulas imply that in the case when $h_{kj} \neq 0$, the operator $h_{kj}^{-1}\mathbf{s}_{kj}$ admits right factorization.

Let us introduce the operator $\mathbf{S}_T$ as follows

$$\mathbf{S}_T \begin{bmatrix} a_1\left(t\right) \\ \vdots \\ a_m\left(t\right) \\ b\left(t\right) \end{bmatrix} = \begin{bmatrix} \varphi_1\left(0, t\right) \\ \vdots \\ \varphi_m\left(0, t\right) \\ \varphi_{m+1}\left(0, t\right) \end{bmatrix}.$$

This operator will be called the transmission operator. From the representations (8), we determine that

$$\mathbf{S}_T = \mathbf{I} + \begin{bmatrix} \mathbf{A}_{11-} & \cdots & \mathbf{A}_{1m-} & \mathbf{A}_{(1,m+1)+} \\ \mathbf{A}_{21-} & \cdots & \mathbf{A}_{2m-} & \mathbf{A}_{(2,m+1)+} \\ \vdots & & \vdots & \vdots \\ \mathbf{A}_{(m+1,1)-} & \cdots & \mathbf{A}_{(m+1,m)-} & \mathbf{A}_{(m+1,m+1)+} \end{bmatrix}.$$

The Volterra properties of the elements of the scattering operator enable us to determine the operators A_{ij-}, $A_{(m+1,j)-}$, $A_{(i,m+1)+}$ for $i, j = 1, 2, .., m$ and $A_{(m+1,m+1)+}$ in terms of the scattering operator's elements.

Proposition 6.1 *The transmission operator* $\mathbf{S}_T$ *can be determined by the scattering operator* $\mathbf{S}_H$ *for the system (11) on the semi-axis.*

3.3 Inverse Scattering Problem on the Half-Plane

We understand the inverse scattering problem as finding the potential $[\sigma, Q](x, t)$ $(x \geq 0, -\infty < t < +\infty)$ by using the known scattering operator $\mathbf{S}_H$ for the system (11) on the semi-axis.

The ISP for the system (11) on the semi-axis is reduced to the ISP for the system on the whole axis through a change of variables $x \to t$, $t \to x$ in the system (11) on the semi-axis with the additional condition that the coefficients are equal to zero for $x < 0$. This can be derived by comparing the formula (6) for the scattering operator on the entire axis with the formula for the scattering problem on the semi-axis.

Proposition 6.2 *The transmission operator* $\mathbf{S}_T$ *is invertible and the inverse* $\mathbf{S}_T^{-1}$ *is the scattering operator for the system obtained by the change of variables* $x \to t$, $t \to x$ *in the system (11) on the half-axis with the additional condition on the coefficients that they are equal to zero for* $x < 0$.

The ISP for such type of systems on the whole axis is solved in Sect. 2 of this paper. Thus, we obtain the following result about the ISP for the system (11) on the semi-axis.

Theorem 6.5 *Let the coefficients of the system (11) be measurable, complex-valued, and square-summable functions. Let* $\mathbf{S}_H$ *be the scattering operator for the system (11) on the semi-axis, with* $\det H \neq 0$. *Then the coefficients of the system (11) are uniquely determined by the scattering operator* $\mathbf{S}_H$.

Remark 6.1 In the general case $\xi_1 = \cdots = \xi_p > 0 > \xi_{p+1} = \cdots = \xi_n$ the system (1) is called the Dirac-type system (called also AKNS system). These systems are of great interest themselves and as auxiliary systems for the integrable nonlinear equations. A procedure to integrate the multidimensional nonlinear evolution equations associated with these linear scattering problems via the GLM equatiion can be found in [12, 15]. The inverse scattering problem for general nonstationary Dirac-type systems (1) on the semi-axis remains a subject for future study.

References

1. Ayoubi, I.S., Nelson, P.: Numerical solution of the inverse scattering problem for hyperbolic systems of n components in semi-infinite media. In: Invariant Imbedding and Inverse Problems, pp.187–208, SIAM, Philadelphia, PA (1990)
2. Åberg, I., Kristensson, G., Wall, J.N.: Propagation of transient electromagnetic waves in time-varying media-direct and inverse scattering problems. Inverse Prob. **11**, 29–49 (1995)
3. Belishev, M.I., Blagovestchenskii, A., Ivanov, S.A.: The two-velocity dynamical system: boundary control of waves and inverse problems. Wave Motion **25**, 83–107 (1997)
4. Bui, D.D., Nelson, P.: Existence and uniqueness of an inverse problem for a hyperbolic system of lidar probing. Inverse Prob. **8**, 821–829 (1992)
5. Burridge, R.: The Gelfand-Levitan, the Marchenko, and the Gopinath-Sondhi integral equations of inverse scattering theory, regarded in the context of inverse impulse-response problems. Wave Motion **2**, 305–23 (1980)
6. Gohberg, I., Krein, M.: Theory and Applications of Volterra Operators in Hilbert Space. American Mathematical Society, Providence, RI (1970)
7. Ismailov, M.I.: Inverse scattering problem for nonstationary Dirac-type systems on the plane. J. Math. Anal. Appl. **365**, 498–509 (2010)
8. Ismailov, M.I.: Inverse scattering problem for the nonstationary Dirac equation on the half-plane. J. Inverse Ill-Posed Prob. (2015)
9. Ismailov, M.I.: Inverse scattering problem for hyperbolic system on a semi-axis in the case of equal number of incident and scattered waves. Inverse Prob. **22**, 955–974 (2006)
10. Iskenderov, N.S., Ismailov, M.I.: Inverse scattering problem for nonstationary Dirac-type systems on the half-plane. J. Diff. Eqn. **246**, 277–290 (2009)
11. Iskenderov, N.S.: Inverse non-stationary scattering problem for hyperbolic system of N equations of the first order on a semi-axis UKR. Math. J. **43**, 1524–32 (1991)
12. Iskenderov, N.S., Ismailov, M.I.: On the inverse scattering transform of a nonlinear evolution equation with 2+ 1 dimensions related to nonstrict hyperbolic systems. Nonlinearity **25**(7), 1967–1979 (2012)
13. Nizhnik, L.P.: Inverse Scattering Problems for Hyperbolic Equations. Nauk, Dumka, Kiev (1973) (in Russian)
14. Nizhnik, L.P.: Inverse Scattering Problems for Hyperbolic Equations. Nauk, Dumka, Kiev (1991) (in Russian)
15. Nizhnik, L.P.: The inverse scattering problem for hyperbolic equations and their application to nonlinear integrable systems. Rep. Math Phys. **26**(2), 261–283 (1988)
16. Nizhnik, L.P., Tarasov, V.G.: The inverse nonstationary scattering problem for a hyperbolic system of equations. Soviet Math. Dokl. **18**(2), 397–401 (1977); Also: Dokl. Akad. Nauk SSSR **233**(3), 300–303 (1977) (in Russian)
17. Nizhnik, L.P., Iskenderov, N.S.: Inverse nonstationary scattering problem for hyperbolic system of three equations of the first order on a semi-axis. Ukrainian Math. J. **42**(7), 825–832 (1990)
18. Shabat, A.B.: An one-dimensional scattering theory. I. Diff. Uravn. **8**, 164–178 (1972). (in Russian)
19. Sung, L.Y., Fokas, A.S.: Inverse problem for $N \times N$ hyperbolic systems on the plane and the $N-$wave interactions. Commun. Pure Appl. Math. **64**, 535–571 (1991)

Quantitative Homogenisation for Differential Equations with Highly Anisotropic Partially Degenerating Coefficients

Shane Cooper and Ilia Kamotski

Abstract We consider a non-uniformly elliptic second-order differential operator with periodic coefficients that models composite media consisting of highly anisotropic cylindrical fibres periodically distributed in an isotropic background. The degree of anisotropy is related to the period of the coefficients via a 'critical' high-contrast scaling. In particular, ellipticity is lost in certain directions as the period, ε, tends to zero. Our primary interest is in the asymptotic behaviour of the resolvent of this operator in the limit of small ε. Two-scale resolvent convergence results were established for such operators in Cherednichenko et al. in (Proc Royal Soc Edinb: Sect A Math 136(1):87–114, 2006, [2]). In this work, we provide an asymptotic description of the resolvent and establish operator-type error estimates. Our approach adopts the general scheme of Cooper et al. (Quantitative multiscale operator-type approximations for asymptotically degenerating spectral problems, [4]) However, we face new challenges such as a directional dependence on the loss of ellipticity in addition to a key 'spectral gap' assumption of the above article only holding in a weaker sense. This results in additional 'interfacial' boundary layer analysis in the vicinity of each fibre to arrive at order-ε operator-type error estimates.

Keywords Quantitative high-contrast homogenisation · Error estimates · Periodic differential operators with rapidly oscillating degenerating coefficients

MSC2020 35B27 · 35J35 · 35A15 · 78M35

1 Introduction

In this work, we analyse the asymptotic behaviour of the solution $u_\varepsilon \in H^1(\mathbb{R}^3)$ to the problem

S. Cooper · I. Kamotski (✉)
University College London, London, UK
e-mail: i.kamotski@ucl.ac.uk

M. Chatzakou et al. (eds.), *Direct and Inverse Problems with Applications*,
Trends in Mathematics 12, https://doi.org/10.1007/978-3-031-98645-1_7

$$\left.\begin{aligned} &-\operatorname{div}\big(A_\varepsilon(\tfrac{x}{\varepsilon})\nabla u_\varepsilon(x)\big)+u_\varepsilon(x)=f(x) \text{ for } f\in L^2(\mathbb{R}^3),\\ &A_\varepsilon(y)=\chi_1(y)I+\chi_0(y)\operatorname{diag}\big(\varepsilon^2,\varepsilon^2,a(y_3)\big), \end{aligned}\right\} \tag{1}$$

for small (period) parameter $0<\varepsilon<1$. Here, $\square := \big[-\frac{1}{2},\frac{1}{2}\big)^3$ is the periodic reference cell of A_ε, $\chi_1 = 1-\chi_0$ and χ_0 is the $\square$-periodic extension of the characteristic function of the cylindrical set $F := B\times\mathbb{R}$ where B is the disk of radius $r<1/2$ centred at the origin. The 1-periodic coefficient a satisfies the usual well-posedness assumptions, i.e. there exists a positive constant $\nu>0$ such that

$$v < a(s) < v^{-1}, \qquad \forall s\in[0,1).$$

Such equations are one possible model for highly anisotropic periodic composite media with partial degeneracies [7].

It was shown in [2] that the solution u_ε to (1) two-scale converges, as $\varepsilon\to 0$, to $u(x)+v(x,y')$ the solution of the two-scale homogenised limit problem

$$\left.\begin{aligned} &-\operatorname{div}A^{\mathrm{h}}\nabla u - a^{\mathrm{h}}\partial_{x_3}^2\big(|B|u+\langle v\rangle\big)+u+\langle v\rangle = f, && x\in\mathbb{R}^3,\\ &-\Delta_{y'}v - a^{\mathrm{h}}\partial_{x_3}^2(u+v)+u+v=f, && x\in\mathbb{R}^3,\ y'\in B,\\ &v(x,y')=0, && x\in\mathbb{R}^3,\ y'\in\partial B, \end{aligned}\right\} \tag{2}$$

where $\langle v\rangle(x) := \int_B v(x,y')\,\mathrm{d}y'$ and A^{h} and a^{h} are given constant coefficients (defined below). Note that the two-scale limit problem is defined on a subspace of $L^2(\mathbb{R}^3\times\square)$ that is larger than the space on which the original problem is posed. As such, one has convergence in the two-scale topology rather than the standard topology, cf. the seminal works [1, 6, 9].

In our recent work [4], we provide a general methodology to upgrade such two-scale convergence results to operator-type error estimates in the space on which the original problem is posed. We begin by the standard procedure of quantitative homogenisation for periodic operators in the whole space to arrive at a family of equivalent problems on the unit cell $\square$. Namely, we sequentially apply, to u_ε, the unitary rescaling $\Gamma_\varepsilon : L^2(\mathbb{R}^3)\to L^2(\mathbb{R}^3)$, $\Gamma_\varepsilon f(x)=\varepsilon^{3/2}f(\varepsilon x)$, then the unitary Gelfand transform $U : L^2(\mathbb{R}^3)\to L^2(\square^*\times\square)$, for $\square^* = [-\pi,\pi)^3$ the dual cell and U being the continuous extension of the mapping

$$Uf(\theta,y) = (2\pi)^{-3/2}\sum_{m\in\mathbb{Z}^3} f(y+m)e^{-\imath\theta\cdot(y+m)}.$$

Then $u_{\varepsilon,\theta}(y) := \big(U\Gamma_\varepsilon u\big)(\theta,y)$ solves

$$-\big(\operatorname{div}+\imath\theta\cdot\big)\varepsilon^{-2}A_\varepsilon(\nabla+\imath\theta)u_{\varepsilon,\theta}+u_{\varepsilon,\theta}=f_{\varepsilon,\theta} \quad y\in\square,\theta\in\square^*,$$

where $f_{\varepsilon,\theta}(y) := \big(U\Gamma_\varepsilon f\big)(\theta,y)\in L^2(\square^*\times\square)$. Following [4], we rewrite this problem as: for each $0<\varepsilon<1$ and $\theta\in\square^*$,

$$\left.\begin{array}{l}\text{Find } u_{\varepsilon,\theta} \in H := H^1_{per}(\square) \text{ the solution to} \\ \varepsilon^{-2} a_\theta(u_{\varepsilon,\theta}, \tilde{u}) + b_\theta(u_{\varepsilon,\theta}, \tilde{u}) = c(f_{\varepsilon,\theta}, \tilde{u}), \quad \forall \tilde{u} \in H\end{array}\right\}, \tag{3}$$

for families of non-negative sesquilinear forms $a_\theta, b_\theta : H \times H \to \mathbb{C}$ and $c : L^2(\square) \times L^2(\square) \to \mathbb{C}$ given by the quadratic forms

$$a_\theta(u, u) =: a_\theta[u] = \int_{\square \setminus F} |(\nabla + \imath\theta)u(y)|^2 \,\mathrm{d}y + \int_F a(y_3)|(\partial_3 + \imath\theta_3)u(y)|^2 \,\mathrm{d}y; \tag{4}$$

$$b_\theta(u, u) =: b_\theta[u] = \int_F |(\nabla' + \imath\theta')u|^2 + \int_\square |u|^2, \quad \nabla' = (\partial_1, \partial_2), \ \theta' = (\theta_1, \theta_2), \tag{5}$$

and

$$c[u] = \int_\square |u|^2.$$

Notice that, $a_\theta + b_\theta$ are a (uniformly equivalent) family of inner products on H and so (3) is well-posed.

The goal now, is to provide leading-order asymptotics for solutions to the family of variational problems (3) that are uniform in θ, ε and the right-hand-side $f_{\varepsilon,\theta}$. Then one can invert the transforms U and Γ_ε applied above to find uniform estimates for the problem (1). In [4], such asymptotics were provided for variational families of the form (3) with non-negative forms a_θ that are (point-wise) coercive in H (after projecting away from the kernel of a_θ). Amongst other things, this ensured that first order correction of $u_{\varepsilon,\theta}$ lived in H which is crucial to the analysis in [4]. Another feature of [4] is that the coercivity constant degenerates quadratically in θ.

Here, the form a_θ (cf. 4) is clearly not coercive (up to the kernel) in $H^1_{per}(\square)$. Indeed, consider

$$V = \{v \in H^1_{per}(\square) : v \text{ constant in } \square \setminus F \text{ and } v \text{ independent of } y_3 \text{ in } F\}. \tag{6}$$

Clearly, a_θ is zero on V when $\theta_3 = 0$. Now, denote the closed subset W of $H^1_{per}(\square)$ that is the orthogonal complement of V with respect to the inner product $a_0 + b_0$ (which is an equivalent H^1 inner product). Then, it is easy to see that a_θ is positive on W. The key assumption of [4] (up to some details) was that a_θ is bounded from below by b_θ on W. This 'spectral-gap' assumption, which holds for many examples of interest, cf. [4, 5], fails in the present example (i.e. for forms 4 and 5). However, one has the following weaker coercivity estimate: there exists a positive constant γ such that

$$a_\theta[w] \geq \gamma c[w], \quad \forall w \in W, \ \forall \theta \in \square^*.$$

Overcoming this absence of full coercivity is one of the main challenges of this current work as it leads to new analysis not present in [4], in particular, we need to investigate internal boundary-layer effects near each fibre.

Another difficultly is related to the directional dependence of the elliptic degeneracy, which is a feature in various highly anisotropic high-contrast composite media (see for example [3]), and the failure of the quadratic degeneracy assumption of [4]. Loosely speaking, in the present context, this assumption would be satisfied if one could demonstrate that $a_\theta[u]$ is bounded from below by $|\theta|^2 c[u]$ for all u. This is clearly not the case. However, for (4), one has the following 'directionally-dependent' quadratic condition: there exists a positive constant γ^* such that

$$a_\theta[u] \geq \gamma^*\Big(|\theta_3|^2 c[\chi_0 u] + |\theta|^2 c[\chi_1 u]\Big), \quad \forall u \in H^1_{per}(\square), \forall \theta \in \square^*.$$

Such a weaking the quadratic gap assumption of [4] present significant challenges. These challenges are too significant to achieve results on the same level of generality as in our previous work [4]. However, we were able to succeed for problem (1) by exploiting problem-specific features.

2 Results

Our first result concerns uniform asymptotics of (3) with operator-norm error estimates. Upon representing V, cf. (6), as the direct sum $V = Z_0 \dot{+} Z_1$, where

$$Z_0 := \mathbb{C}, \ Z_1 := \{v \in H^1_{per}(\square) \ : \ v = 0 \text{ in } \square \backslash F \text{ and } v \text{ independent of } y_3 \text{ in } F\},$$

we introduce the following problem on V:

$$\left.\begin{aligned} &\text{Find } (z_0, z_1) \in Z_0 \times Z_1 \text{ the solution to} \\ &\varepsilon^{-2} A^{\rm h}\theta \cdot \theta \, z_0 \overline{\tilde{z}_0} + a^{\rm h}\varepsilon^{-2}|\theta_3|^2 \big(z_0 + z_1, \tilde{z}_0 + \tilde{z}_1\big)_{L^2(B)} \\ &\qquad + b_0(z_0 + z_1, \tilde{z}_0 + \tilde{z}_1) = c(\mathcal{E}_\theta f_{\varepsilon,\theta}, \tilde{z}_0 + \tilde{z}_1) \quad \forall \tilde{z}_0 \in Z_0, \tilde{z}_1 \in Z_1 \end{aligned}\right\}.$$

Here, $\mathcal{E}_\theta : L^2(\square) \to L^2(\square)$ is the unitary operator given by multiplication by the function $\exp(\imath\theta' \cdot y')$, $a^{\rm h} := \left(\int_0^1 a^{-1}\right)^{-1}$ is the one-dimensional homogenised matrix and $A^{\rm h}$ is the constant, symmetric 3×3 perforated-fibre homogenised matrix given, in [2], as

$$A^{\rm h} := |\square \backslash F| e_3 \otimes e_3 + \sum_{\alpha,\beta=1}^{2} e_\alpha \otimes e_\beta \int_{\square' \backslash B} (\partial_\beta N_\alpha + \delta_{\alpha\beta}),$$

for $N_\alpha : H^1_{per}(\square' \backslash B) \to \mathbb{R}$, $\square' := \big[-\frac{1}{2}, \frac{1}{2}\big)^2$, $\alpha = 1, 2$, that solve the 'cell problem'

$$\left.\begin{aligned}&\int_{\Box'\backslash B}(\nabla' N_\alpha + e_\alpha)\nabla'\phi = 0, \quad \forall \phi \in H^1_{per}(\Box'\backslash B)\\ &\int_{\Box'\backslash B} N_\alpha = 0\end{aligned}\right\}.$$

Here, e_k, $k = 1, 2, 3$, are the canonical Euclidean basis vectors.

Our first main result is the following uniform estimate:

$$\|u_{\varepsilon,\theta} - \mathcal{E}^*_\theta(z_0 + z_1)\|_{L^2(\Box)} \lesssim \varepsilon \| f_{\varepsilon,\theta}\|_{L^2(\Box)},$$

which can be rewritten in operator-theoretic language as follows.

Theorem 7.1 *There exists a positive constant C such that the inequality*

$$\|(\mathcal{L}_{\varepsilon,\theta} + I)^{-1} - \mathcal{E}^*_\theta(\mathbb{L}_{\theta/\varepsilon} + I)^{-1} P_0 \mathcal{E}_\theta\|_{L^2(\Box)\to L^2(\Box)} \le C\varepsilon$$

holds for all $0 < \varepsilon < 1$ *and* $\theta \in \Box^*$.

The operators in the above theorem are the following: for each ε and θ, $(\mathcal{L}_{\varepsilon,\theta} + I)$ is the unbounded self-adjoint operator in $L^2(\Box)$ generated by problem (3); now denote $\mathcal{H}_0$ to be the closure of V in $L^2(\Box)$ and $P_0 : L^2(\Box) \to \mathcal{H}_0$ the orthogonal projection; then, finally, for each $\xi \in \mathbb{R}^3$, $(\mathbb{L}_\xi + I)$ is the unbounded self-adjoint operator in $\mathcal{H}_0$ generated by the inner product

$$\begin{aligned}\mathbb{S}_\xi(z_0 + z_1, \tilde{z}_0 + \tilde{z}_1) := A^{\rm h}\xi \cdot \xi z_0\overline{\tilde{z}_0} + a^{\rm h}|\xi_3|^2\big(z_0 + z_1, \tilde{z}_0 + \tilde{z}_1\big)_{L^2(B)}\\ + b_0(z_0 + z_1, \tilde{z}_0 + \tilde{z_1}).\end{aligned}$$

Each of the non-negative operators $\mathcal{L}_{\varepsilon,\theta}$ and $\mathbb{L}_\xi$ have compact resolvents. Consequently, Theorem 7.1 readily implies uniform approximations for the associated eigenvalues. Indeed, for $\{\lambda^{(k)}_\varepsilon(\theta)\}_{k=1}^\infty$ and $\{\Lambda^{(k)}(\xi)\}_{k=1}^\infty$ denoting the eigenvalues of $\mathcal{L}_{\varepsilon,\theta}$ and $\mathbb{L}_\xi$ respectively, labelled in increasing order and repeated according to multiplicity, one has the following result.

Theorem 7.2 *For each $k \in \mathbb{N}$, there exists a constant $C_k > 0$ such that*

$$\left|\lambda^{(k)}_\varepsilon(\theta) - \Lambda^{(k)}\left(\tfrac{\theta}{\varepsilon}\right)\right| \le C_k\varepsilon, \quad \forall \theta \in \Box^*, \ \forall\, 0 < \varepsilon < 1.$$

Theorem 7.1 enables one to approximate the solution of the original problem (1) in terms of the two-scale homogenized limit problem (2). To this end, let us define the Hilbert space

$$\mathcal{H} := L^2(\mathbb{R}^3; \mathcal{H}_0),$$

$\mathcal{P} : L^2(\mathbb{R}^3 \times \Box) \to \mathcal{H}$ the orthogonal projection, and define the dense subspace

$$\mathcal{V} := H^1(\mathbb{R}^3) \dot{+} \mathcal{Z}_1, \quad \mathcal{Z}_1 := \{v \in L^2(\mathbb{R}^3, Z_1) \,:\, \partial_{x_3} v \in L^2(\mathbb{R}^3 \times \Box)\},$$

of $\mathcal{H}$. Finally, let $\mathcal{L}_0 + I$ be the unbounded self-adjoint operator defined by the closed quadratic form

$$Q[u+v] = \int_{\mathbb{R}^3} \Big(A^{\mathrm{h}} \nabla u \cdot \overline{\nabla u} + a^{\mathrm{h}} \|\partial_{x_3}(u+v)\|^2_{L^2(B)} + b_0[u+v] \Big) \, \mathrm{d}x,$$

with form domain $\mathcal{V}$.

To provide error estimates in $\mathcal{L}(L^2(\mathbb{R}^3))$ between the resolvents of $\mathcal{L}_\varepsilon$ and $\mathcal{L}_0$, one needs to address the fact that the operators are defined on different function spaces. We overcome this issue via a 'two-scale interpolation' operator $J_\varepsilon : L^2(\mathbb{R}^3) \to L^2(\mathbb{R}^3 \times \square)$, introduced in [4] (see [8] for a similar object) given by

$$J_\varepsilon := \Gamma_\varepsilon^{-1} \mathcal{F}^{-1} \chi \mathcal{E}_\theta^* U \Gamma_\varepsilon.$$

Here, $\chi : L^2(\square^* \times \square) \to L^2(\mathbb{R}^3 \times \square)$ extends functions by zero outside of $\square^*$; $\mathcal{F}$ is the Fourier transform acting on the first variable (and Γ_ε, Γ_ε^{-1} are understood to act on the first variable). It was established in [4] that J_ε is almost-unitary in the following sense:

$$J_\varepsilon^* J_\varepsilon = I, \quad \text{and} \quad J_\varepsilon J_\varepsilon^* \to I \text{ strongly as } \varepsilon \to 0.$$

Then, the following result holds.

Theorem 7.3 *There exists a constant* $K > 0$ *such that the inequality*

$$\|(\mathcal{L}_\varepsilon + I)^{-1} - J_\varepsilon^* (\mathcal{L}_0 + I)^{-1} \mathcal{P} J_\varepsilon \|_{\mathcal{L}(L^2(\mathbb{R}^3))} \le K\varepsilon,$$

holds for all $0 < \varepsilon < 1$.

Our results imply comparisons between the spectral properties of $\mathcal{L}_\varepsilon$ and $\mathcal{L}_0$. For example, upon noting the fibre decomposition $\big(\mathcal{F}\mathcal{L}_0\mathcal{F}^{-1}\big)(\xi) = \mathbb{L}_\xi$, one sees that Theorem 7.2 provides uniform estimates between the spectral band functions of the operators $\mathcal{L}_\varepsilon$ and $\mathcal{L}_0$.

The proof of the announced results will appear elsewhere.

Acknowledgements The authors would like to thank Prof. Valery Smyshlyaev for bringing this problem to our attention.

References

1. Allaire, G.: Homogenization and two-scale convergence. SIAM J. Math. Anal. **23**(6), 1482–1518 (1992)
2. Cherednichenko, K.D., Smyshlyaev, V.P., Zhikov, V.V.: Non-local homogenized limits for composite media with highly anisotropic periodic fibres. Proc. Royal Soc. Edinb.: Sect. A Math. **136**(1), 87–114 (2006)
3. Cooper, S.: Quasi-periodic two-scale homogenisation and effective spatial dispersion in high-contrast media. Calc. Var. PDEs. **57**(76) (2018)
4. Cooper, S., Kamotski, I., Smyshlyaev, V.P.: Quantitative multiscale operator-type approximations for asymptotically degenerating spectral problems. Preprint https://arxiv.org/abs/2307.13151
5. Kamotski, I., Smyshlyaev, V.P.: Two-scale homogenization for a general class of high contrast PDE systems with periodic coefficients. Appl. Anal. **98**(1), 64–90 (2019)
6. Nguetseng, G.: A general convergence result for a functional related to the theory of homogenization. SIAM J. Math. Anal. **20**(3), 608–623 (1989)
7. Smyshlyaev, V.P.: Propagation and localization of elastic waves in highly anisotropic periodic composites via two-scale homogenization. Mech. Mater. **41**(4), 434–447 (2009)
8. Wellander, N.: The two-scale Fourier transform approach to homogenization; periodic homogenization in Fourier space. Asymptotic Anal. **62**(12), 1–40 (2009)
9. Zhikov, V.V.: On an extension of the method of two-scale convergence and its applications. Sbornik: Math. **191**(7), 973–1014 (2000)

Rigidity Results for Evolution PDEs on Homogeneous Lie Groups

Alessia E. Kogoj

Abstract We extend a result obtained by Suzuki and Watson in 2001 for caloric functions: under certain conditions, a volume mean value identity characterizes heat balls. We obtain an inverse mean value property of solutions of a class of evolution equations of Hörmander-type.

Keywords Degenerate parabolic equations · Kolmogorov-type operators · Hypoelliptic operators · Rigidity properties · Inverse problems

MSC2020 35H10 · 35K65 · 35K70 · 53C24 · 35K05 · 35B99.

1 Introduction: The Heat Operator Case

The aim of this work is to extend a theorem by Suzuki and Watson concerning an inverse property for caloric functions. In order to recall the Suzuki–Watson Theorem, we need to introduce some notations and results.

Let

$$\mathcal{H} = \Delta_{\mathbb{R}^N} - \partial_t$$

be the heat operator in $\mathbb{R}^{N+1} = \mathbb{R}^N \times \mathbb{R} \ni z = (x, t)$. Let G be the Gauss–Weierstrass kernel:

$$G(x,t) = \frac{1}{(4\pi t)^{\frac{N}{2}}} \exp\left(-\frac{|x|^2}{4t}\right) \text{ if } t > 0, \qquad G(x,t) = 0 \text{ if } \ t \le 0.$$

A. E. Kogoj (✉)
Department of Pure and Applied Sciences, Università di Urbino Carlo Bo, Urbino, PU, Italy
e-mail: alessia.kogoj@uniurb.it

M. Chatzakou et al. (eds.), *Direct and Inverse Problems with Applications*, Trends in Mathematics 12, https://doi.org/10.1007/978-3-031-98645-1_8

Let O be an open subset of $\mathbb{R}^{N+1}$. We say that $u \in C^\infty(O, \mathbb{R})$ is caloric in O if and only if

$$\mathcal{H}u = 0 \text{ in } O.$$

We define the heat balls or caloric balls centred at $z_0 \in \mathbb{R}^{N+1}$ with radius $r > 0$ as follows:

$$\Omega_r(z_0) = \left\{ z \in \mathbb{R}^{N+1} : G(z_0 - z) > \frac{1}{r} \right\}.$$

The caloric functions satisfy the following analogous of the Gauss Mean Value property.

Theorem 1 (Pini 1951; Watson 1973) *Let $u \in C(O, \mathbb{R})$. Then, $u \in C^\infty(O, \mathbb{R})$ and $\mathcal{H}u = 0$ in O if and only if*

$$u(z_0) = M_r(u)(z_0) := \frac{1}{r} \int_{\Omega_r(z_0)} u(\zeta) W(z_0 - \zeta)\, d\zeta,$$

for every heat ball $\Omega_r(z_0)$ such that $\overline{\Omega_r(z_0)} \subseteq O$.
Here, W stands for

$$W(\zeta) = W(\xi, \tau) = \frac{|\xi|^2}{4\tau^2}.$$

As a consequence, since $\zeta \longmapsto G(\zeta - z)$ is caloric in $\Omega_r(z_0)$ for every $z \notin \Omega_r(z_0)$, one has

$$G(z_0 - z) = \frac{1}{r} \int_{\Omega_r(z_0)} G(\zeta - z) W(z_0 - \zeta) d\zeta,$$

for every heat ball $\Omega_r(z_0)$, $\overline{\Omega_r(z_0)} \subseteq O$, and for every $z \notin \Omega_r(z_0)$.
Suzuki and Watson proved that this identity is actually a rigidity property of the heat balls.

Theorem 2 (Suzuki and Watson 2001) *Let $z_0 = (x_0, t_0) \in \mathbb{R}^{N+1}$ and let D be a bounded open set of $\mathbb{R}^{N+1}$. Assume that for a suitable constant $c > 0$,*

$$\int_D G(\zeta - z) W(z_0 - \zeta)\, d\zeta = c\, G(z_o - z) \quad \forall z \notin D.$$

Assume also that

$$\zeta \longmapsto (\chi_D - \chi_{\Omega_c(z_0)})(\zeta) W(z_0 - \zeta) \in L^p \text{ for some } p > \frac{n}{2} + 1,$$

then,

$$D = \Omega_c(z_0).$$

Here and henceforth, χ_E denotes the characteristic function of a set E in $\mathbb{R}^{N+1}$.

We observe that Suzuki–Watson method is inspired by the one introduced by Shapiro to study quadrature domains for harmonic functions [9]. A careful analysis of such a method shows that it only uses ideas and results from the classical Potential Analysis for the heat equation. All the basic ingredients of Suzuki–Watson method are today present in Potential Analysis for wide classes of linear second order hypoelliptic evolution Partial Differential Equations.

Recently, with Lanconelli in [4], we extended the Suzuki–Watson Theorem to the prototype of Kolmogorov operators. We are currently extending the same method to more general classes. More precisely, we are considering the heat operators on Carnot groups and the evolution operators of Hörmander-type that we will describe in the third section. In this way we also extend previous results obtained with Lanconelli and Tralli in 2014 [6].

2 The Kolmogorov Operators Case

We consider operators in $\mathbb{R}^{N+1}$ of the following type

$$\mathcal{K} = \operatorname{div}(A\nabla) + \langle Bx, \nabla\rangle - \partial_t,$$

where div, ∇, $\langle\cdot,\cdot\rangle$ stand respectively for the divergence, the Euclidean gradient and the inner product in $\mathbb{R}^N$. $A = (a_{ij})_{i,j=1,\dots,N}$ and $B = (b_{ij})_{i,j=1,\dots,N}$ are $N \times N$ matrices with real constant coefficients, A is symmetric and non-negative definite. Moreover, the matrix A has the following block structure:

$$A = \begin{bmatrix} A_0 & 0 \\ 0 & 0 \end{bmatrix},$$

where A_0 is a $p_0 \times p_0$ symmetric and strictly positive definite matrix, $p_0 \le N$. If $p_0 < N$, i.e. if $\mathcal{K}$ is a degenerate elliptic-parabolic operator, the matrix B has to take the form

$$B = \begin{bmatrix} 0 & 0 & 0 & 0 & 0 \\ B_1 & 0 & 0 & 0 & 0 \\ 0 & B_2 & \dots & 0 & 0 \\ \vdots & \vdots & \ddots & \vdots & \vdots \\ 0 & 0 & \dots & B_n & 0 \end{bmatrix},$$

where B_j is a $p_j \times p_{j-1}$ matrix with maximum rank p_j; $j = 1, 2, \dots, n$, $p_0 \ge p_1 \ge \dots \ge p_n \ge 1$ and $p_0 + p_1 + \dots + p_n = N$.

These assumptions guarantee the hypoellipticity of the operator (other conditions equivalent can be found in [1], see also [8]), the left translation invariance and homogeneity of degree two with respect to the Lie group $\mathbb{K} = (\mathbb{R}^{N+1}, \cdot, \delta_\lambda)$ with composition law

$$(x,t)\cdot(x',t') = (x' + E(t')x, t+t').$$

E represents the matrix

$$E(s) := e^{-sB}, \quad s \in \mathbb{R},$$

and the group of dilations is

$$\delta_\lambda : \mathbb{R}^{N+1} \longrightarrow \mathbb{R}^{N+1}, \quad \delta_\lambda(x,t) := (\lambda x^{(p_0)}, \lambda^3 x^{(p_1)}, \ldots, \lambda^{2n+1} x^{(p_n)}, \lambda^2 t);$$

where $x^{(p_i)} \in \mathbb{R}^{p_i}$, $i = 0, \ldots, n$, and $\lambda > 0$. An explicit fundamental solution for $\mathcal{K}$ is given by

$$\Gamma(z,\zeta) := \gamma\left(\zeta^{-1}\cdot z\right) \text{ for } z, \zeta \in \mathbb{R}^{n+1},$$

where $\zeta^{-1} = (\xi,\tau)^{-1} = (-E(-\tau)\xi, -\tau)$ is the opposite of ζ with respect to the composition law and

$$\gamma(z) = \gamma(x,t) := \begin{cases} 0 & \text{for } t \leq 0 \\ \frac{(4\pi)^{-N/2}}{\sqrt{\det C(t)}} \exp\left(-\frac{1}{4}\left\langle C^{-1}(t)\, x, x\right\rangle\right) & \text{for } t > 0 \end{cases}.$$

The natural number $Q = p_0 + 3p_1 + \cdots + (2r+1)p_n + 2$ is the homogeneous dimension related to the family of dilations δ_λ. The matrix $C(t)$ is defined by the following integral strictly positive definite for every $t > 0$

$$C(t) = \int_0^t E(s) A E^T(s)\, ds$$

A celebrated example of an operator belonging to this class is the operator introduced in 1934 by Kolmogorov in studying diffusion phenomena from a probabilistic point of view. Let $\mathbb{I}_n$ be the identity $n \times n$ matrix, let N be equal to $2n$ and let the matrices A and B be of the type

$$A = \begin{pmatrix} \mathbb{I}_n & 0 \\ 0 & 0 \end{pmatrix} \quad \text{and} \quad B = \begin{pmatrix} 0 & 0 \\ \mathbb{I}_n & 0 \end{pmatrix}.$$

We recover the kinetic Kolmogorov operator

$$\operatorname{div}(A\nabla) + \langle Bx, \nabla\rangle - \partial_t = \sum_{i=1}^n \partial^2_{x_i} + \sum_{i=1}^n x_i \partial_{x_{n+i}} - \partial_t \ \text{ in } \mathbb{R}^{2n+1}.$$

We define now the $\mathcal{K}$-balls centred at z_0 and radius $r > 0$

$$\Omega_r(z_0) = \left\{ z \in \mathbb{R}^{N+1} : \Gamma(z_0, z) > \frac{1}{r} \right\},$$

and over these balls we define the kernel W as follows

$$W(z) = W(x,t) := \frac{\langle AC^{-1}(t)x, C^{-1}(t)x\rangle}{4}.$$

We observe that W extends the Pini–Watson kernel of the heat operator. Also in this more general setting we have a representation formula for the $\mathcal{K}$-harmonic functions in O, i.e., the solutions to $\mathcal{K}u = 0$ in O.

Theorem 3 (Kupcov 1983) *Let $u \in C(O,\mathbb{R})$. Then, $u \in C^\infty(O,\mathbb{R})$ and $\mathcal{K}u = 0$ in O if and only if*

$$u(z_0) = M_r(u)(z_0) =: \frac{1}{r}\int_{\Omega_r(z_0)} u(\zeta)W(z_0^{-1}\circ\zeta)\,d\zeta,$$

for every $\mathcal{K}$-ball $\Omega_r(z_0)$ such that $\overline{\Omega_r(z_0)} \subseteq O$.

We prove that this property characterizes the $\mathcal{K}$-balls, extending to the Kolmogorov setting the Suzuki–Watson theorem.

Theorem 4 (Theorem 1.1, [4]) *Let $z_0 \in \mathbb{R}^{N+1}$ and let D be a bounded open subset of $\mathbb{R}^{N+1}$ such that, for a suitable $r > 0$,*

$$\int_D \Gamma(\zeta, z)W(z_0^{-1}\circ\zeta)\,d\zeta = r\Gamma(z_0, z) \quad \forall z \in \mathbb{R}^{N+1}\setminus D.$$

If, moreover,

$$\zeta \longmapsto (\chi_D - \chi_{\Omega_r(z_0)})W(z_0^{-1}\circ\zeta) \in L^p \text{ for some } p > \frac{Q}{2},$$

then $D = \Omega_r(z_0)$.

An idea of the key steps of the proof can be found in [2]. We refer to [4] for complete details.

3 A Class of Evolution Equations of Hörmander-Type

We consider a class of linear second order operators of the following type

$$\mathcal{L} = \sum_{j=1}^m X_j^2 + X_0 - \partial_t = \sum_{j=1}^m X_j^2 + Y \qquad \text{in } \mathbb{R}^{N+1}.$$

The X_j' s are linear first order differential operators in $\mathbb{R}^N$ with coefficients of class C^∞. We assume that the following conditions are satisfied.

(H1) There exists a homogeneous Lie group in $\mathbb{R}^{N+1}$,

$$\mathbb{L} = (\mathbb{R}^{N+1}, \circ, d_\lambda)$$

such that

(i) $X_1, \dots, X_m, Y$ are left translation invariant on $\mathbb{L}$.
(ii) $X_1, \dots, X_m$ are d_λ-homogeneous of degree one and Y is d_λ-homogeneous of degree two.

(H2) For every $(x, t), (y, \tau) \in \mathbb{R}^{N+1}$, $t > \tau$, there exists an *$\mathcal{L}$-admissible* path $\eta : [0, T] \longrightarrow \mathbb{R}^{N+1}$ such that $\eta(0) = (x, t)$, $\eta(T) = (y, \tau)$. The curve η is called $\mathcal{L}$-admissible if it is absolutely continuous and satisfies

$$\eta'(s) = \sum_{j=1}^{m} \lambda_j(s) X_j(\eta(s)) + \mu(s) Y(\eta(s)) \qquad \text{a.e. in } [0, T]$$

for suitable piecewise constant real functions $\lambda_1, \dots, \lambda_m$, and μ, $\mu \geq 0$.

We remark that conditions (H1) and (H2) imply the hypoellipticity of $\mathcal{L}$.

This class of operators contains, as particular cases, the heat operators on Carnot groups and the operators constructed with the *link procedure* [5]. They were introduced and studied in [3] where it is proved the existence of a global fundamental solution Γ for $\mathcal{L}$, smooth out of the diagonal, non-negative and with support in a halfspace, together with some of its basic properties. Furthermore, a Mean Value Theorem for $\mathcal{L}$-harmonic functions holds. Given $z_0 \in \mathbb{R}^{N+1}$ and $r > 0$ we define the $\mathcal{L}$-ball of centre z_0 and radius r as follows

$$\Omega_r(z_0) := \left\{ z \in \mathbb{R}^{N+1} \ : \ \Gamma(z^{-1} \circ z_0) > \frac{1}{r} \right\}.$$

Theorem 5 (Theorem 2.7, [3]) *Let u be a (smooth) solution to $\mathcal{L}u = 0$ in the open set $O \subseteq \mathbb{R}^{N+1}$. Then, for every $\mathcal{L}$-ball $\Omega_r(z_0)$ with closure contained in O, we have*

$$u(z_0) = \frac{1}{r} \int_{\Omega_r(z_0)} u(\zeta) W(z_0, \zeta) \ d\zeta,$$

where

$$W(z, \zeta) = \frac{|\nabla_{\mathcal{L}} \Gamma(z, \zeta)|^2}{\Gamma^2(z, \zeta)}, \qquad \nabla_{\mathcal{L}} = (X_1, \dots, X_m).$$

Here $\nabla_{\mathcal{L}}$ acts on the variable ζ.

Thanks also to a Poisson–Jensen type representation formula proved in [7], we can extend Suzuki–Watson theorem to this large class of operators adding the assumption that $\mathbb{R}^{N+1} \setminus \overline{\Omega_r(z_0)}$ is *connected to infinity by $\mathcal{L}^*$-trajectories* in the sense of

Definition 1.3 in [6]. Here we observe that in the case that $\mathcal{L}$ coincides with the heat operator on a stratified Lie group this condition is always verified. Under this assumption, we can generalize the Suzuki–Watson theorem even in this setting.

Theorem 6 *Let $z_0 \in \mathbb{R}^{N+1}$ and let D be a bounded open subset of $\mathbb{R}^{N+1}$ such that, for a suitable $r > 0$,*

$$\int_D \Gamma(\zeta, z) W(z_0^{-1} \circ \zeta)\, d\zeta = r\Gamma(z_0, z) \quad \forall z \in \mathbb{R}^{N+1} \setminus D.$$

If, moreover,

$$\zeta \longmapsto (\chi_D - \chi_{\Omega_r(z_0)}) W(z_0^{-1} \circ \zeta) \in L^p \text{ for some } p > \frac{Q}{2},$$

then $D = \Omega_r(z_0)$.
Here Q represents the homogeneous dimension related to the group of dilations (d_λ).

Acknowledgements This work is a summary of the lectures that I delivered at the Summer School "Direct and Inverse problems with applications, and related topics" held at Ghent University in August 2024. It is a great pleasure to thank the organizers, Marianna Chatzakou, Michael Ruzhansky, Karel Van Bockstal, and the whole Ghent Analysis and PDE group for the kindest hospitality and the fruitful and friendly atmosphere.

References

1. Kogoj, A.E.: Liouville-type theorems for Kolmogorov and Ornstein–Uhlenbeck operators. Lect. Notes Sem. Interdiscip. Mat. **15**(2), 45–53 (2020)
2. Kogoj, A.E.: On a rigidity result for Kolmogorov-type operators. Bruno Pini Math. Anal. Sem. **15**(1), 98–111 (2024)
3. Kogoj, A.E., Lanconelli, E.: An invariant Harnack inequality for a class of hypoelliptic ultraparabolic equations. Mediterr. J. Math. **1**, 51–80 (2004)
4. Kogoj, A.E., Lanconelli, E.: A rigidity theorem for Kolmogorov-type operators. Rend. Istit. Mat. Univ. Trieste **57** (2025), Art. No. 5, 16 pages
5. Kogoj, A.E., Lanconelli, E.: Link of groups and homogeneous Hörmander operators. Proc. Am. Math. Soci. **135**, 2019–2030 (2007)
6. Kogoj, A.E., Lanconelli, E., Tralli, G.: An inverse mean value property for evolution equations. Adv. Differ. Equ. **19**(7–8), 783–804 (2014)
7. Kogoj, A.E., Tralli, G.: Blaschke, Privaloff, Reade and Sacks Theorems for diffusion equations on Lie groups. Potential Anal. **38**(4), 1103–1122 (2013)
8. Lanconelli, E., Polidoro, S.: On a class of hypoelliptic evolution operators. Rend. Sem. Mat. Univ. Politec. Torino **52**(1), 29–63 (1994)
9. Shapiro, H.S.: The Schwarz function and its generalization to higher dimensions. Wiley, New York (1992)
10. Suzuki, N., Watson, N.A.: A characterization of heat balls by a mean value property for temperatures. Proc. Am. Math. Soc. **129**, 2709–2714 (2001)

An Overview of Dualities in Non-commutative Harmonic Analysis

Yulia Kuznetsova

Abstract The aim of this mini-course is to give an overview of duality tools of the non-commutative harmonic analysis. By volume limitations, it cannot be complete, but tries to be as wide as possible and accessible to starting Ph.D. students. I start with the unitary dual of a locally compact group, then discuss the Plancherel measure on it and disintegration results. The following outlines of symmetric spaces, non-commutative L^p spaces and quantum groups are very brief, but allow to present several results on Fourier multipliers in non-commutative settings.

Keywords Harmonic analysis · Duality · Unitary dual · Group algebras · Locally compact groups · Multipliers

MSC2020 22D35 · 43A22 · 22E30 · 46L67 · 43A90

1 Introduction

The aim of this article is to give an overview of tools of the non-commutative harmonic analysis.

Most of the general theory presented below can be bound in the books of Dixmier [4] and Gerald Folland [6]. On spherical theory, the recommended reference is Sigurdur Helgason, Groups and Geometric Analysis [8]. And on quantum groups, Thomas Timmermann, An Invitation to Quantum Groups and Duality [19]. Other sources are research articles.

Duality and Fourier transform

Duality is one of the main tools in classical harmonic analysis. The Fourier transform $\mathcal{F}$ takes convolution to pointwise multiplication, thus allowing to simplify differential equations and to estimate their solutions. Keep in mind that solutions of differential equations are convolutions with Green functions, so we are particularly interested in

Y. Kuznetsova (✉)
University of Franche-Comté, Besançon, France
e-mail: ykuznets@univ-fcomte.fr

M. Chatzakou et al. (eds.), *Direct and Inverse Problems with Applications*,
Trends in Mathematics 12, https://doi.org/10.1007/978-3-031-98645-1_9

convolution operators $T : f \mapsto f * k$; the kernel k can be a function, measure, or a distribution.

This theory is fully available on an abelian locally compact group, or in short abelian LCG, G: the group of characters of G

$$\widehat{G} = \{\phi : G \to \mathbb{T} \text{ continuous homomorphism}\}$$

is also an abelian LCG, and $\widehat{\widehat{G}} \simeq G$. The Fourier transform $\mathcal{F}$ satisfies $\mathcal{F}(f * g) = \mathcal{F}(f)\mathcal{F}(g)$ and is isometric on L^2, thus it is easy to say when T is bounded on L^2: exactly when $\mathcal{F}(k)$ is bounded. If G is non-abelian, such a duality is not possible. The first natural analogue of the group of characters is $\widehat{G}$: the set of all unitary irreducible representations of G. But $\widehat{G}$ is no more a group. Three solutions exist:

- Work with $\widehat{G}$ as a topological or a measured space
  $\longrightarrow$ Plancherel measure, leads to an isometry on L^2 in many cases
- Consider only finite-dimensional representations
  $\longrightarrow$ tensor categories, especially useful for compact groups
- Study group algebras
  $\longrightarrow$ Duality of quantum groups

And a special domain is the study of subclasses of operators which admit Fourier-like methods (analysis on symmetric spaces).

I will speak of all of them except for the second one which could not enter in the volume available. A good reference for this approach is [15].

2 Unitary Dual

Abelian groups are well described by their characters having scalar values, and the Fourier transform of a function is again a scalar-valued function. In the non-commutative case, representations act on Hilbert spaces, possibly of infinite dimension, and the Fourier transform will be an operator-valued function on $\widehat{G}$.

Recall that every unitary representation π of G on a Hilbert space H_π generates a representation of $L^1(G)$ by

$$\pi(f) = \int_G f(g)\pi(g)dg; \tag{1}$$

the integral is understood as an operator such that for every $x, y \in H_\pi$

$$\langle \pi(f)x, y\rangle = \int_G f(g)\langle \pi(g)x, y\rangle dg.$$

This representation extends by continuity to $C^*(G)$. Conversely, every *-representation of $C^*(G)$ defines a unitary representation of G (provided that

$\pi(C^*(G))H$ is dense in H). A representation π is called irreducible if it has no nontrivial closed invariant subspaces. If G is abelian, every irreducible π is one-dimensional.

Representations π, ρ of G on Hilbert spaces H_π, H_ρ are called equivalent if there exists a unitary $T : H_\pi \to H_\rho$ such that $T\pi(g) = \rho(g)T$ for every $g \in G$.

Definition 1 $\widehat{G}$ is the set of equivalence classes of *unitary irreducible* representations.

Nilpotent case

It is usually a complicated task to describe $\widehat{G}$. For a nilpotent simply connected Lie group G, a powerful orbit method is available. Examples of such groups include strictly upper-triangular matrices and Heisenberg groups $\mathbb{R} \ltimes \mathbb{R}^n$.

Let $\mathfrak{g}$ denote the Lie algebra of G. At $X \in \mathfrak{g}$, the *adjoint action* of $g \in G$ is

$$Ad(g) : \exp(tX) \mapsto g\exp(tX)g^{-1}.$$

On $\mathfrak{g}^*$, this generates the *coadjoint action*:

$$Ad^*(g) : \lambda \mapsto \lambda \circ Ad(g).$$

We say that $\lambda \sim \mu$ if $\lambda = Ad^*(g)\mu$ for some g. This is an equivalence relation. Without going into further detail, let us state

Theorem 1 (Kirillov [12]) *$\widehat{G}$ is in bijection with $\mathfrak{g}^*/\sim$.*

Heisenberg group

This example will help to see the identification above. We set $\mathbb{H} = \mathbb{R} \ltimes \mathbb{R}^2$, or the group of matrices $\begin{pmatrix} 1 & a & c \\ 0 & 1 & b \\ 0 & 0 & 1 \end{pmatrix}$.

Its Lie algebra is $\mathfrak{g} = \langle X, Y, Z \rangle$ with

$$X = \begin{pmatrix} 0 & 1 & 0 \\ 0 & 0 & 0 \\ 0 & 0 & 0 \end{pmatrix} \quad Y = \begin{pmatrix} 0 & 0 & 0 \\ 0 & 0 & 1 \\ 0 & 0 & 0 \end{pmatrix} \quad Z = \begin{pmatrix} 0 & 0 & 1 \\ 0 & 0 & 0 \\ 0 & 0 & 0 \end{pmatrix},$$

so that $[X, Y] = Z$ and $[X, Z] = [Y, Z] = 0$.

The space $\mathfrak{g}^*$ is 3-dimensional and the coadjoint action (we do not calculate it here) is

$$Ad^*(a, b, c) : (\mu, \nu, \lambda) \mapsto (\mu + b\lambda, \nu - a\lambda, \lambda).$$

We see that the orbit of $(\mu, \nu, 0)$ is only this point, and if $\lambda \neq 0$, the orbit is the plane $\Omega_\lambda = \{(\mu, \nu, \lambda) : (\mu, \nu) \in \mathbb{R}^2\}$. Denote $\Omega = \{\Omega_\lambda : \lambda \in \mathbb{R}\}$.

For $\mu, \nu \in \mathbb{R}$, the one-dimensional representation corresponding to $(\mu, \nu, 0)$ is given by

$$\pi_{\mu,\nu}(a,b,c) = e^{2\pi i(a\mu+b\nu)};$$

for $\lambda \in \mathbb{R}$, infinite-dimensional irreducibles act on $f \in L^2(\mathbb{R})$ as

$$\pi_\lambda(a,b,c)(f)(t) = e^{2\pi i\lambda(c+bt)} f(t+a).$$

By direct calculations, one can verify that for $f \in L^1(\mathbb{H}) \cap L^2(\mathbb{H})$,

$$\|f\|_2^2 = \int_{\mathbb{R}} Tr[\pi_\lambda(f)^*\pi_\lambda(f)]\,|\lambda| d\lambda.$$

The measure

$$\mu(\{\pi_{\mu,\nu}\}) = 0, \quad \mu|_\Omega = |\lambda| d\lambda \tag{2}$$

is called the Plancherel measure on $\widehat{\mathbb{H}}$, and we will discuss it below in the general case.

Topology on the unitary dual
This dual is more often considered as a measure space, but one starts with a topology. For $\pi \in \widehat{G}$, its kernel $\ker \pi$ is an ideal in $C^*(G)$. We define the *closure* of any $E \subset \widehat{G}$:

$$\overline{E} = \{\pi \in \widehat{G} : \pi \equiv 0 \text{ on } I_E = \cap_{\rho\in E} \ker \rho\}.$$

There exists (by Kuratowski theorem) a topology on $\widehat{G}$ in which this is the closure map. One calls it the Fell, or hull-kernel topology. $\widehat{G}$ is not always Hausdorff.

If G is *abelian*, then $C^*(G) \simeq C_0(\widehat{G})$. For $\rho \in \widehat{G}$, we have $\ker \rho = \{f \in C_0(\widehat{G}) : f(\rho) = 0\}$, so that $I_E = \{f \in C_0(\widehat{G}) : f|_E \equiv 0\}$. Then

$$\overline{E} = \{\pi \in \widehat{G} : f(\pi) = 0 \text{ for every } f \in C_0(\widehat{G}) \text{ vanishing on } E\}.$$

Clearly this is the usual closure of E.

2.1 Plancherel Measure

For $f \in L^1(G)$ and $\pi \in \widehat{G}$, set $\widehat{f}(\pi) = \pi(f)$, as given by the formula (1): this defines the noncommutative Fourier transform. Sometimes one sets instead $\pi(f) = \int_G f(g)\pi(g^{-1})dg$.

For groups of so called "Type I", there is a direct analogy with the classical case:

Theorem 2 *Let G be a unimodular Type I group. There exists a unique positive measure μ on $\widehat{G}$ such that for every $f \in L^1(G) \cap L^2(G)$*

$$\|f\|_2^2 = \int_{\widehat{G}} Tr\big(\widehat{f}(\pi)^*\widehat{f}(\pi)\big)d\mu(\pi). \tag{3}$$

There exists a non-unimodular version [6, Theorem 7.50] but the group should be "almost" Type I.

Let us now define the term "Type I". Recall that a subalgebra A in $B(H)$ (the space of bounded operators on a Hilbert space H) is called a von Neumann algebra if it is closed in strong or, equivalently, weak operator topology. This happens if and only if A is equal to its second commutant A''; and for a set $S \subset B(H)$, its commutant is defined as $S' = \{y \in B(H) : yx = xy \ \forall x \in S\}$. Next, a von Neumann algebra A is a factor if $A \cap A' = \mathbb{C}1$.

Now, being Type I means: if $\pi(G)''$ is a factor, then it is of type I, that is, isomorphic to $B(H)$ (of finite or infinite dimension).This is also equivalent to the following: if π is irreducible, then $\pi\big(C^*(G)\big)$ contains all compact operators. If G is not of Type I, then it has necessarily both Type II and Type III representations [7]. An example of a non-Type I group is the free group of two generators.

Other terms for this class are GCR and postliminal groups. These notions have appeared independently and have different definitions, but have turned to be the same classes of groups as Type I.

In relation to the dual, the following facts are known. For any G, if $\pi_1 = \pi_2 \in \widehat{G}$, then clearly $\ker \pi_1 = \ker \pi_2$. For Type I groups, the converse is true. Moreover, a second countable group G is Type I if and only if the map $\pi \mapsto \ker \pi$ is injective, and if and only if the topology of $\widehat{G}$ is T_0.

Finally, if G is Type I and π irreducible, then the von Neumann algebra $\pi(G)''$ admits a trace. Moreover, we know that:

- Every connected semisimple and nilpotent Lie group is Type I.
- A discrete group is Type I if and only if it is almost abelian (that is, has an abelian subgroup of finite index).

Compact groups
In this case, $\widehat{G}$ is discrete: if $\chi_\pi : g \mapsto Tr\,\pi(g)$ is the character of π, then $\rho(\chi_\pi) = 0$ if $\pi \neq \rho$ and $\pi(\chi_\pi) \neq 0$, so that $\widehat{G} \setminus \{\pi\}$ is closed, thus $\{\pi\}$ open.

The Plancherel theorem takes the following form:

$$\|f\|_2^2 = \sum_{\pi \in \widehat{G}} d_\pi Tr\big(\pi(f)\pi(f)^*\big).$$

Inverse Fourier transform
The Plancherel identity implies that for $f, g \in L^1 \cap L^2(G)$

$$\begin{aligned}\langle f, g\rangle &= \int_{\widehat{G}} Tr\big(\pi(g)^*\pi(f)\big)d\mu(\pi)\\ &= \int_{\widehat{G}}\int_G Tr\big(\overline{g(x)}\pi(x)^*\pi(f)\big)dx\,d\mu(\pi),\end{aligned}$$

so that

$$f(x) = \int_{\widehat{G}} Tr\big(\pi(x)^*\pi(f)\big)d\mu(\pi).$$

This allows in particular to define pseudo-differential operators on such groups [17].

Plancherel measure: the support

The support of μ (intersection of all closed sets of full measure) is the set of $\rho \in \widehat{G}$ *weakly contained* in λ, that is, such that for every $\xi \in H_\rho$, $\varepsilon > 0$, and compact $K \subset G$ there exist $f_k \in L^2(G)$ with

$$\sup_{g\in K} \left|\langle \rho(g)\xi, \xi\rangle - \sum_{k=1}^{n} \tilde{f}_k * f_k(g)\right| < \varepsilon.$$

Here $\tilde{f}(x) = \bar{f}(x^{-1})$.

G is amenable iff $\operatorname{supp}\mu = \widehat{G}$. This is not the case, for example, for $SL_2(\mathbb{R})$.

Example: Heisenberg group

Recall that $\widehat{\mathbb{H}} = \{\pi_{\mu,\nu}\} \cup \{\pi_\lambda\}$, and only π_λ enter in the Plancherel formula (2). For $k \in L^1(\mathbb{H}) \cap L^2(\mathbb{H})$,

$$\begin{aligned}\pi_\lambda(k)f(t) &= \int_H k(a,b,c)e^{2\pi i\lambda(c+bt)} f(t+a)dadbdc \\ &= \int_{\mathbb{R}} \mathcal{F}_2\mathcal{F}_3(k)(x-t, -\lambda, -\lambda t)f(x)dx\end{aligned}$$

so that $\widehat{k}(\pi_\lambda)$ is an integral operator with the kernel as above.

Plancherel measure: non-unimodular case

The Plancherel measure still exists, with Formulas (3) and (4) valid, but the group should be "almost Type I" [6, Theorem 7.50] and we have to adjust the definition of the Fourier transform setting

$$\widehat{f}(\pi) = \pi(f)D_\pi,$$

with $\pi(f)$ as in (1). Here D_π are self-adjoint operators on H_π such that $D_\pi\pi(g) = \Delta(g)^{1/2}\pi(g)D_\pi$ for every $g \in G$. Their existence is not obvious, but becomes clearer knowing that every π can be realized on the space $H_\pi = L^2(G, V_\pi)$ where V_π is also a Hilbert space; one sets then $D_\pi f(g) = \Delta(g)^{-1/2} f(g)$ for $f \in H_\pi$ and $g \in G$.

The Plancherel identity for $f, g \in L^1 \cap L^2(G)$ becomes

$$\langle f, g\rangle = \int_{\widehat{G}}\int_G Tr\big(D_\pi\overline{g(x)}\pi(x)^*\pi(f)D_\pi\big)dxd\mu(\pi),$$

and the inverse Fourier transform

$$f(x) = \int_{\widehat{G}} Tr\big(\pi(x)^*\pi(f)D_\pi^2\big)d\mu(\pi).$$

"ax+b" group
This is the group of matrices

$$\begin{pmatrix} a & b \\ 0 & 1 \end{pmatrix}$$

with $a > 0$, $b \in \mathbb{R}$, which correspond to the affine maps on the real line: $g_{a,b}x = ax + b$.

We could consider also $G_n = \{(a, b) : a \in \mathbb{R}, b \in \mathbb{R}^n\}$, acting the same way. If we denote the above group by G, then G is non-unimodular: the left Haar measure is $\frac{1}{a^2}dadb$, the right one $\frac{1}{a}dadb$, the modular function is a^{-1}.

One can describe $\widehat{G}$ by the orbit method, arriving at: $\widehat{G} = \{\pi_\lambda : \lambda \in \mathbb{R}\} \cup \{\pi^+, \pi^-\}$, acting as

$$\pi_\lambda(a, b) = a^{i\lambda};$$

π^+, π^- act on $H^+ = L^2(0, +\infty)$ and $H^- = L^2(-\infty, 0)$ respectively by the same formula:

$$\pi^\pm(a, b)\phi(t) = \sqrt{a}\, e^{2\pi ibt}\phi(at).$$

The Plancherel measure μ on $\widehat{G}$ is such that $\mu(\{\pi_\lambda\}) = 0$ for every λ, and

$$\|f\|_2^2 = Tr\big[\widehat{f}(\pi^+)^*\widehat{f}(\pi^+)\big] + Tr\big[\widehat{f}(\pi^-)^*\widehat{f}(\pi^-)\big]$$

for $f \in L^2(G)$. The adjustment operators are $D^\pm\phi(s) = \sqrt{|s|}\phi(s)$, and $\widehat{f}(\pi^\pm) = \pi^\pm(f)D^\pm$. One verifies directly that $\widehat{f}(\pi^+)$ is an integral operator with the kernel

$$k_f^+(s, t) = \frac{\sqrt{s}}{t}\mathcal{F}_2 f(s^{-1}t, -s),$$

the Fourier transform being applied in the second variable only. A similar formula holds for $\widehat{f}(\pi^-)$.

2.2 Disintegration

The Plancherel measure has much more properties than just the isometry of the Fourier transform on $L^2(G)$. The next property to discuss is the decomposition of representations into direct integrals of irreducibles [4].

Let (Ω, μ) be a measured space and H_ω a Hilbert space for every $\omega \in \Omega$. A *measurable field* of Hilbert spaces is a linear subspace $\Gamma \subset \prod_{\omega\in\Omega} H_\omega$ such that

- $\omega \mapsto \|x(\omega)\|$ is measurable for every $x \in \Gamma$;
- there exist $(x_n) \subset \Gamma$ such that $\{x_n(\omega)\}_{n\in\mathbb{N}}$ is total in H_ω for every ω;
- if $\omega \mapsto \langle x(\omega), y(\omega)\rangle$ is measurable for every $x \in \Gamma$, then $x \in \Gamma$.

We denote

$$\Gamma := \int^{\oplus} H_\omega d\mu(\omega).$$

If $A(\omega) \in B(H_\omega)$ for every ω and $(A(\omega)x(\omega)) \in \Gamma$ for every $x \in \Gamma$, then $A = \int^{\oplus} A(\omega)d\mu(\omega)$ is said to be a mesurable field of operators on Γ. It is bounded if and only if

$$\|A\| = \text{esssup}_{\omega\in\Omega}\|A(\omega)\| < \infty.$$

Disintegration of representations
If π_ω is a unitary representation of G on H_ω for every $\omega \in \Omega$ and $(\pi_\omega(g))$ is a measurable field of operators for every $g \in G$, then $\pi = \displaystyle\int^{\oplus} \pi_\omega d\mu(\omega)$, the direct integral of π_ω, is a representation of G on $\int^{\oplus} H_\omega$. Conversely, if G is a second countable locally compact group, then every unitary representation π of G on a separable H is unitarily equivalent to a direct integral $\displaystyle\int^{\oplus} \pi_\omega d\mu(\omega)$ where π_ω are irreducible for almost all ω [4, 8.5.2], [6, 7.37].

This disintegration is not unique, and this can be seen in the example of the most important representation: of the regular one. The left regular representation λ of G acts at $f \in L^2(G)$ as

$$[\lambda(g)f](h) = f(g^{-1}h).$$

A non-uniqueness example [21]. Let $\mathbb{F}_2$ be the free group of two generators a and b. Then λ has two decompositions into irreducibles

$$\lambda = \int_{\mathbb{T}} U_\theta d\theta = \int_{\mathbb{T}} V_\theta d\theta$$

such that none of U_θ, $V_{\theta'}$ are equivalent.

We introduce briefly these representations. Let G_a, G_b denote all words in $\mathbb{F}_2$ which end by a, a^{-1} or b, b^{-1} respectively, and set $G_a' = G_a \cup \{e\}$, $G_b' = G_b \cup \{e\}$. For $\theta \in \mathbb{T}$, the space H_θ is the closed linear span of the indicator functions $\{\delta_g : g \in G_b'\}$ which are its orthonormal basis; U_θ acts as

$$U_\theta(a)\delta_e = \theta\delta_e, \quad U_\theta(a)\delta_g = \delta_{ag}, \ g \in G_b;$$
$$U_\theta(b)\delta_g = \delta_{bg}.$$

For V_θ, the construction is the same with a and b interchanged.
One can show that U_θ, $V_{\theta'}$ are irreducible and pairwise inequivalent.

Disintergation over the dual
To achieve uniqueness, or otherwise stated, to integrate over the dual, one has to return to Type I groups. Multiplicities occur in this process:

Theorem 3 ([6, 7.40]) *Let G be a second countable Type I group, σ a unitary representation of G on a separable H. Then there exist finite pairwise orthogonal measures μ_n, $n \in \mathbb{N} \cup \{\infty\}$, on $\widehat{G}$, such that*

$$\sigma \simeq \int^{\oplus} \pi d\mu_1(\pi) \oplus 2 \int^{\oplus} \pi d\mu_2(\pi) \oplus \cdots \oplus |\mathbb{N}| \int^{\oplus} \pi d\mu_\infty(\pi).$$

These measures are unique up to equivalence.

In particular, for the regular representation

$$\lambda = \int_{\pi \in \widehat{G}} \pi \otimes I_{\bar{H}_\pi} \, d\mu(\pi) \tag{4}$$

with the Plancherel measure μ; the group should be second countable and Type I [6, 7.44, 7.50]. If G is compact, we recover the familiar decomposition

$$\lambda = \bigoplus_{\pi \in \widehat{G}} d_\pi \pi.$$

Translation-invariant operators

Not only $\lambda(g)$ but every translation invariant operator admits such a disintegration. Recall that convolution operators are translation invariant; often, converse is true [20], [5, Corollary 3.2.1]. If T is bounded on L^2, then:

$$T \text{ is left invariant} \Leftrightarrow T \in \lambda(G)'',$$

where S'' is the bicommutant in $B(L^2(G)))$, or in other terms, the von Neumann algebra generated by S.

In the same assumptions on G as above, for every $T \in \lambda(G)''$

$$T = \int_{\widehat{G}}^{\oplus} T_\pi \otimes I_{\bar{H}_\pi} \, d\mu(\pi). \tag{5}$$

Only the support of λ enters in this decomposition. If $Tf = k * f$, then

$$T_\pi = \widehat{k}(\pi) = \int_G k(x)\pi(x)dx.$$

Heisenberg group: sub-Laplacian

One can verify that, on $L^2(\mathbb{R})$, for every $\lambda \in \mathbb{R}$

$$\pi_\lambda(X) = \sqrt{|\lambda|}\frac{\partial}{\partial t}, \quad \pi_\lambda(Y) = i \,\mathrm{sgn}\lambda\sqrt{|\lambda|}\, t, \quad \pi_\lambda(Z) = 2\pi i \lambda.$$

If $\mathcal{L} = X^2 + Y^2$ is the sub-Laplacian, then, even if it is unbounded, the disintegration (5) holds for it [5], with

$$\mathcal{L}_{\pi_\lambda} f(t) = |\lambda|\big(f''(t) - t^2 f(t)\big). \tag{6}$$

3 Symmetric Spaces

Let G be a semisimple Lie group, K its maximal compact subgroup. By Iwasawa decomposition, $G = KAN$ with $A \simeq \mathbb{R}^l$ abelian, N nilpotent. Let $\mathfrak{a} \simeq \mathbb{R}^l$ and $\mathfrak{g}$ denote the Lie algebras of A and G.

The principal series representations $\pi_{\sigma,\lambda}$ of G are parametrized by irreducibles $\sigma \in \widehat{K}$ and $\lambda \in \mathfrak{a}^*$; if $\sigma = 1$, they have K-invariant vectors and are called spherical. One can realize them on subspaces of $L^2(K)$.

For every $\lambda \in \mathfrak{a}^*$, the function ϕ_λ on G is the following coefficient of $\pi_{1,\lambda}$:

$$\phi_\lambda(x) = \langle \pi_\lambda(x)1, 1\rangle_{L^2(K)}.$$

For every $x \in G$ one can write uniquely $x = k_x a_x n_x$; set $H(x) = \log a_x$, then

$$\phi_\lambda(x) = \int_K e^{-(i\lambda+\rho)\left(H(x^{-1}k)\right)} dk,$$

where $\rho \in \mathfrak{a}^*$ is defined by the algebraic structure of $\mathfrak{g}$.

Spherical transform
For f on G, we define the spherical transform

$$\mathcal{H}(f)(\lambda) = \int_G f(x)\phi_\lambda(x^{-1})dx,$$

so that $\mathcal{H}(f)$ is a function on $\mathfrak{a}^*$. It has many similarities to the classical Fourier transform: For f, g in the space $C^\infty(K \setminus G/K)$ of K-biinvariant smooth functions on G, we have

- $\mathcal{H}(f * g) = \mathcal{H}(f)\mathcal{H}(g)$;
- $\|\mathcal{H}(f)\|_2 = \|f\|_2$;
- inversion formula: $f(x) = \int_{\mathfrak{a}^*} \mathcal{H}(f)(\lambda)\phi_\lambda(x)|\boldsymbol{c}(\lambda)|^{-2} d\lambda$,

where $\boldsymbol{c}$ is the important Harish-Chandra $\boldsymbol{c}$-function on G.

Spherical functions and differential operators
Every ϕ_λ is an eigenfunction of all differential operators on $C^\infty(G)$ invariant with respect to G on the left and K on the right. In particular, the structure of a Riemannian manifold on G/K gives rise to its Laplace-Beltrami operator $\mathcal{L}$ which can be considered as an operator on $C^\infty(G)$ in the above class. It is convenient to denote $\mathcal{L}_\rho = \mathcal{L} + |\rho|^2$, then

$$\mathcal{L}_\rho \phi_\lambda = |\lambda|^2 \phi_\lambda.$$

For every function of $\mathcal{L}_\rho$, we have also $F(\mathcal{L}_\rho)\phi_\lambda = F(|\lambda|^2)\,\phi_\lambda$. Being left invariant, it is a convolution operator $f \mapsto f * k_F$, and the inversion formula for the spherical transform implies that

$$k_F(x) = \int_{\mathfrak{a}^*} F(|\lambda|^2)\,\phi_\lambda(x)|\boldsymbol{c}(\lambda)|^{-2}d\lambda.$$

Fourier analysis methods then apply, since this is just an integral over $\mathfrak{a}^* \simeq \mathbb{R}^l$. The complicated part is the spherical function, for which one seeks simpler approximations; it is known for example, that for "good" $x = e^H \in A$ and $\|H\| \to \infty$,

$$\phi_\lambda(e^H) \sim \sum_{k=1}^{n} \boldsymbol{c}(s_k\lambda)e^{(is_k\lambda-\rho)(H)},$$

where $\{s_k\}$ form a certain group of reflections on $\mathfrak{a}$, called the Weyl group.

Examples

(1) If $G = SL_n(\mathbb{R})$, then $K = SO_n(\mathbb{R})$,
$A = \{$ diagonal matrices in G with positive diagonal $\}$,
$N = \{$ upper-triangular matrices with diagonal 1 $\}$, and
$G/K \simeq AN = \{$ upper-triangular matrices in G with positive diagonal $\}$.

(2) The "ax+b" group can be identified with G/K where $G = SO(2,1)_o$ is the connected component of the identity of $SO(2,1)$, and $K = SO(2)$.

4 Traces and Weights

A *weight* on a C^* or von Neumann algebra A is a map $\tau : A^+ \to [0, +\infty]$ such that

- τ is additive,
- $\tau(tx) = t\,\tau(x)$ for $t \in \mathbb{R}_+$, $x \in A^+$, with $0 \cdot (+\infty) = 0$.
  If in addition $\tau(xx^*) = \tau(x^*x)$ for all $x \in A$, then τ is a *trace*.

A trace τ is *faithful* if $\tau(x) = 0$ only for $x = 0$, and *semifinite* if for any $x \in A^+$ there exists $y \in A^+$ such that $y \leq x$ and $\tau(y) < \infty$.

Moreover we know that:

- On $A = B(H)$: the usual trace Tr.
- On $A = C_0(G)$: $\tau(f) = \int_G f$, so a weight can be viewed as a non-commutative integral.

Plancherel weight/trace On $A = C_r^*(G)$ or $VN(G)$, we can set $\tau(L_k) = k(e)$ if $k = f^* * f$, or equivalently

$$\tau(L_f^* L_f) = \tau(L_{f^* * f}) = \|f\|_2^2.$$

If G is unimodular, this is a trace. In general, just a weight.

We can see that the equality (3) is in fact a disintegration of $\tau(L_f^* L_f)$ into $\tau_\pi\big(\widehat{f}(\pi)^* \widehat{f}(\pi)\big)$ where every τ_π is the usual trace.

Example
Consider again the Heisenberg group $\mathbb{H}$ and the sub-Laplacian $\mathcal{L} = X^2 + Y^2$ on it. For any $u \in \mathbb{R}$, by (3)

$$\tau\big(I_{(0,u)}(\mathcal{L})\big) = \int_{\mathbb{R}} \tau_\lambda\big(I_{(0,u)}(\mathcal{L}_{\pi_\lambda})\big)|\lambda| d\lambda;$$

the operator $\mathcal{L}_{\pi_\lambda}$ is given by (6) and its eigenvalues are known, so one can show [1] that

$$\tau\big(I_{(0,u)}(\mathcal{L})\big) = \int_{\mathbb{R}} \#\{k : s_{k,\lambda} < u\}|\lambda| d\lambda \lesssim u^{3/2}. \tag{7}$$

Non-commutative L_p spaces
Let M be a von Neumann algebra with a faithful semifinite normal trace τ (normal means that $\tau(\sup x_i) = \sup \tau(x_i)$ for every bounded increasing net in A^+). For $x \in M$, define

$$\|x\|_p = \tau(|x|^p)^{1/p}$$

and $L_p(M)$ as the completion of $\{x \in M : \|x\|_p < \infty\}$.

For $p = \infty$, $\|\cdot\|_\infty$ is the usual norm of $M = L^\infty(M)$.

In particular, we know that:

1. If $M = L^\infty(X)$ with $\tau(f) = \int_X f d\tau$, then $L^p(M) = L^p(X)$.
2. If $M = \lambda(G)'' = VN(G)$, then with the Plancherel weight, for $f \in L^2(G)$

$$\|L_f\|_{L^2(M)} = \tau(L_f^* L_f)^{1/2} = \|f\|_2.$$

Weak L_p spaces
One can define *generalized singular numbers* $\mu_t(x)$ for $x \in M$, $t \in \mathbb{R}_+$:

$$\mu_t(x) = \inf\{\lambda > 0 : \tau\big(I_{(|\lambda|,+\infty)}(x)\big) < t\}$$

with the help of spectral projections of x, and set

$$\|x\|_{p,\infty} = \sup_{t>0} t^{1/p} \mu_t(x).$$

The obtained completed space $L^{p,\infty}(M)$ is called the non-commutative weak L_p space, or Lorentz space.

For a decreasing function $f \geq 0$ on $\mathbb{R}_+$, we get $\mu_t(f) = f(t)$. If x is a compact operator on a Hilbert space, then $\mu_n(x) = \lambda_n(|x|)$ are the eigenvalues of $|x| = (x^*x)^{1/2}$.

Non-unimodular case

If G is not unimodular, the Plancherel weight is not a trace. Non-commutative L^p spaces can be defined still, with complex interpolation available for them; but the Lorentz spaces not, it is an open problem.

There are three definitions of non-tracial L^p spaces, each of them with its advantages and disadvantages: by Haagerup (exposed by Terp [18]), Izumi [10] and Hilsum [9]. We will not go further here.

Quantum groups

This notion generalizes both $C_0(G)$ and $C_r^*(G)$, and provides a duality between them. An equivalent theory is built upon von Neumann algebras rather than C^*-algebras, and includes $L^\infty(G)$ and $VN(G)$. Many new examples, not coming from groups, are included in this theory.

Definition 2 (*Kustermans and Vaes* [13]) A *locally compact quantum group* (LCQG) is a von Neumann algebra M with a map $\Delta : M \to M \otimes M$, called comultiplication, and two weights ϕ, ψ on M such that

- Δ is a unital normal $*$-homomorphism;
- Δ is coassociative: $(1 \otimes \Delta)\Delta = (\Delta \otimes 1)\Delta$;
- ϕ, ψ are normal and semifinite;
- ϕ is left invariant and ψ is right invariant: for $\omega \in M_*^+$ and $x \in M^+$, if $\phi(x)$ or $\psi(x)$ respectively is finite, then

$$\phi\big((\omega \otimes 1)\Delta(x)\big) = \phi(x)\omega(1),$$
$$\psi\big((1 \otimes \omega)\Delta(x)\big) = \psi(x)\omega(1).$$

On $M = L^\infty(G)$, the weights are integrals with respect to the left and right Haar measures. On $M = L^\infty(G)$, they are Plancherel weights.

Duality: There exists a dual LCQG $(\widehat{M}, \widehat{\Delta}, \widehat{\phi}, \widehat{\psi})$ such that $\widehat{\widehat{M}} = M$. The algebras $L^\infty(G)$ and $VN(G)$ are duals of each other. There are also analogues of the Fourier transform: $*$-homomorphisms with dense ranges

$$\lambda : M_* \to \widehat{M}, \quad \widehat{\lambda} : \widehat{M}_* \to M.$$

For $M = L^\infty(G)$, we have $M_* = L^1(G)$ and $\lambda : f \mapsto L_f \in VN(G)$. The map λ is thus the left regular representation of $L^1(G)$. If G is abelian, then $VN(G)$ is identified with $L^\infty(\widehat{G})$ via the usual Fourier transform, with $\widehat{L_f g} = \widehat{f * g} = \widehat{f}\widehat{g}$, so modulo this identification L_f is the operator of multiplication by $\widehat{f}$ and thus deserves its name as the Fourier transform of f.

For $1 \le p \le 2$, there are L_p-Fourier transforms (Cooney [3], Caspers [2]): $\mathcal{F}_p$ maps $L^p(M)$ to $L^{p'}(\widehat{M})$, with $1/p + 1/p' = 1$. In the case of $M = L^\infty(G)$, we have $L^p(M) = L^p(G)$, and $\mathcal{F}_p : f \mapsto L_f \Delta^{1/p'} \in L^{p'}(VN(G))$, where we need to multiply first by a power of the modular function.

M is said of Kac type is ϕ, ψ are traces. The example above shows that in the non-Kac case, in particular for a non-unimodular group, $\mathcal{F}_p$ are different operators for different p, as opposed to the classical case where only their domains are different.

Hausdorff-Young inequality
If G is an abelian locally compact group, then for $1 \le p \le 2$ $\|\widehat{f}\|_{p'} \le C_p \|f\|_p$. On quantum groups this is also true [2]:

$$\|\mathcal{F}_p(f)\|_{L^{p'}(\widehat{M})} \le \|f\|_{L^p(M)},$$

with an equality for $p = 2$. In the Kac case, a stronger inequality holds [22], involving the weak norm $\|f\|_{L^{p,p'}(M)}$.

5 Multipliers

In the classical case, a multiplier with a symbol m is the operator $T_m = \mathcal{F}^{-1}(m\widehat{f})$. On a quantum group of Kac type, one can adopt the same approach: for $m \in \widehat{M}$, we set

$$T_m(y) = \mathcal{F}^{-1}\big(m\mathcal{F}(y)\big),$$

$y \in M$. We assume from now that both M and $\widehat{M}$ are of Kac type. Up to my knowledge, there is no theory of non-Kac multipliers, as those would require a coordination of $\mathcal{F}_p$ transforms for different p. An exception is given by spherical multipliers on symmetric spaces: these groups are non-unimodular, but the multipliers in question are described by the spherical transform which reduces to a commutative convolution subalgebra of functions on G/K.

Say that T_m is an $L^p - L^q$ *Fourier multiplier* if it extends to a bounded map from $L^p(M)$ to $L^q(M)$.

Theorem 4 (Zhang [22]) *For $1 < p \le 2 \le q < \infty$ and r such that $1/r = 1/p - 1/q$, every $m \in L^{r,\infty}(\widehat{M})$ is an $L^p - L^q 5$ Fourier multiplier with*

$$\|T_m\|_{L^p(M)\to L^q(M)} \le C_{p,q}\, \|m\|_{L^{r,\infty}(\widehat{M})}.$$

Example
The case of $M = L^\infty(G)$ with a unimodular group G was considered before by Akylzhanov and Ruzhansky [1]. They give also the following example: on the Heisenberg group $\mathbb{H}$ with the sub-Laplacian $\mathcal{L} = X^2 + Y^2$,

$$\|(1-\mathcal{L})^{-s}\|_{L^{r,\infty}(VN(\mathbb{H}))} \lesssim \sup_{u>0} u^{3/2r}(1+u)^{-s},$$

which is finite as soon as $s \geq \frac{3}{2r}$, so that in these assumptions $(1-\mathcal{L})^{-s}$ is bounded from $L^p(\mathbb{H})$ to $L^q(\mathbb{H})$. The estimate of the weak L^r norm is based on (7).

In the dual case of $M = VN(G)$ on a unimodular group G, the symbol of a multiplier is a bounded function $m : G \to \mathbb{C}$. The map

$$T_m : L_f \mapsto L_{mf}$$

has the above form: here $\mathcal{F} : L_f \mapsto f$ is the Fourier transform of M, which is the inverse to that of $\widehat{M} = L^\infty(G)$, so that $L_{mf} = \mathcal{F}^{-1}(m\mathcal{F}(f))$.

Free group of n ***generators*** $G = \mathbb{F}_n$
Let $|g|$ be the length of the reduced word $g \in G$, and

$$m_t(g) = e^{-t|g|}.$$

For positivity reasons, T_{m_t} is bounded on $L^p(VN(G))$ for every $p \in [1, +\infty]$. We can calculate the weak L^r norm of m_t.

We have $|m_t(g)| \geq \alpha$ iff $|g| \leq -\frac{1}{t}\log\alpha = \frac{1}{t}|\log\alpha|$ since $m_t < 1$.
The next formula is valid for $0 < \alpha < 1$, otherwise $\lambda_m(\alpha) = 0$:

$$\lambda_m(\alpha) = |\{g \in G : |m(g)| > \alpha\}| \leq (2n)^{t^{-1}|\log\alpha|}.$$

$$\|m_t\|_{L^{r,\infty}(L^\infty(G))} = \sup_{\alpha>0} \alpha\lambda_m^{1/r}(\alpha) \leq \sup_{0<\alpha<1} \alpha^{1-(tr)^{-1}\log(2n)}$$

and this is finite iff $1-(tr)^{-1}\log(2n) \geq 0$, that is, iff $t \geq \frac{\log(2n)}{r}$. Under this condition and p, q as in Theorem 4, T_m is an $L^p - L^q$ multiplier.

Even more is true, on any discrete group (Zhang [22]): if $|\xi| \leq m_t$ and $1 < p < \infty$, $\frac{1}{p*} = |\frac{1}{2} - \frac{1}{p}|$, then $\|T_\xi\|_{L^p(VN(G))} \lesssim \|\xi\|_{L^{p*,\infty}(G)} \leq \|m_t\|_{L^{p*,\infty}(G)}$.

By different methods, another estimate was obtained by Junge et al. [11]: For $1 < p \leq q < \infty$, $\|T_{m_t}\|_{L^p(VN(G))\to L^q(VN(G))} \leq 1$ as soon as $t > C\log\frac{q-1}{p-1}$.

For L^p ***multipliers***, a major result on $\mathbb{R}^d$ is

Theorem 5 (Hörmander–Mikhlin 1956–1960) *If m is $\frac{d}{2}+1$ times differentiable and with some $C_m > 0$, with usual notations for $\alpha \in \mathbb{N}^d$,*

$$\max_{0\leq|\alpha|\leq[\frac{d}{2}]+1} \sup_{\xi\in\mathbb{R}^d} |\xi|^{|\alpha|}|\partial^\alpha m(\xi)| \leq C_m, \tag{8}$$

then m is an $L^p(\mathbb{R}^d)$ multiplier for every $1 < p < \infty$.

Briefly, the estimate depends on a finite number of derivatives of m which should decrease polynomially. It is applicable in particular to $e^{-|\xi|}$ or to $(1+|\xi|^2)^{-r}$ with

r large enough. It turns out that similar assumptions have sense for multipliers on group algebras.

Free group $\mathbb{F}_\infty$ of a countable set of generators
Let $\{g_k\}_{k\in\mathbb{N}}$ be the generators of G. Let m be a function on $\mathbb{R}^d$, satisfying (8). For $g = g_{i_1}^{k_1} \dots g_{i_n}^{k_n}$, set

$$\tilde{m}(g) = \begin{cases} m(k_1, \dots, k_d) & \text{if } n \leq d; \\ m(k_1, \dots, k_n, 0, \dots, 0) & \text{if } n < d. \end{cases}$$

Then $T_{\tilde{m}}$ is bounded on $L^p(VN(\mathbb{F}_\infty))$ (Mei et al. [14]).

Multipliers on the group algebra of $SL_n(\mathbb{R})$
For $g \in SL_n(\mathbb{R})$, set $|g| := \max(\|g - e\|, \|g^{-1} - e\|)$. Let $X_1, \dots, X_{n^2-1}$ be the standard basis of

$$\mathfrak{sl}_n(\mathbb{R}) = \{A \in M_n(\mathbb{R}) : Tr\, A = 0\}.$$

For $\gamma = (j_1, \dots, j_{|\gamma|})$ and a function m denote

$$d_g^\gamma m(g) = X_{j_1} \dots X_{j_{|\gamma|}} m.$$

Theorem 6 (Parcet et al. [16]) *If for all $|\gamma| \leq [\frac{n^2}{2}] + 1$ and $g \in G$*

$$|g|^{|\gamma|} |d_g^\gamma m(g)| \leq C_m,$$

then T_m is completely bounded on $L^p(VN(G))$ for $1 < p < \infty$.

These examples close our brief survey.

References

1. Akylzhanov, R., Ruzhansky, M.: $L_p - L_q$ multipliers on locally compact groups. J. Funct. Anal. **278**(3), 108324 (2020)
2. Caspers, M.: The l_p-fourier transform on locally compact quantum groups. J. Oper. Theory **69**(1), 161–193 (2013)
3. Cooney, T.: A hausdorff-young inequality for locally compact quantum groups. Int. J. Math. **21**(12), 1619–1632 (2010)
4. Dixmier, J.: C^*-Algebras. North-Holland (1969)
5. Fischer, V., Ruzhansky, M., Quantization on Nilpotent Lie Groups. Birkhäuser (2016)
6. Folland, G.: A Course in Abstract Harmonic Analysis. CRC Press (1995)
7. Glimm, J.: Type i c^*-algebras. Ann. Math. **73**(3), 572–612 (1961)
8. Helgason, S.: Groups and Geometric Analysis. AMS (1984)
9. Hilsum, M.: Les espaces l^p d'une algèbre de von neumann définies par la derivée spatiale. J. Funct. Anal. **40**, 151–169 (1981)
10. Izumi, H.: Constructions of non-commutative l^p-spaces with a complex parameter arising from modular actions. Int. J. Math. **8**, 1029–1066 (1997)

11. Junge, M., Palazuelos, C., Parcet, J., Perrin, M.: Hypercontractivity in group von Neumann algebras. Am. Math. Soc. **249**(1183) (2014)
12. Kirillov, A.A.: Lectures on the Orbit Method. AMS (2004)
13. Kustermans, J., Vaes, S.: Locally compact quantum groups. Ann. Sci. Ecole Norm. Sup. (4) **33**(6), 837–934 (2000)
14. Mei, T., Ricard, E., Xu, Q.: A Mikhlin multiplier theory for free groups and amalgamated free products of von Neumann algebras. Adv. Math. **403**, Paper No. 108394, 32 (2022)
15. Neshveyev, S., Tuset, L.: Compact Quantum Groups and Their Representation Categories. SMF, Cours Spécialisés (2013)
16. Parcet, J., Ricard, E., de la Salle, M.: Fourier multipliers in $sl_n(\searrow)$. Duke Math. J. **171**, 1235–1297 (2022)
17. Ruzhansky, M., Turunen, V.: Pseudo-Differential Operators and Symmetries. BirkhäUser (2010)
18. Terp, M.: L^p spaces associated with von Neumann algebras. Notes, Report No. 3a + 3b, Kobenhavns Universitets Matematiske Institut (1981)
19. Timmermann, Th.: An Invitation to Quantum Groups and Duality. EMS (2008)
20. Wendel, J.G.: Left centralizers and isomorphisms of group algebras. Pac. J. Math. **2**, 251–261 (1952)
21. Yoshizawa, H.: Some remarks on unitary representations of the free group. Osaka Math. J. **3**(1), 55–63 (1951)
22. Zhang, H.: $l_p - l_q$ fourier multipliers on locally compact quantum groups. J. Fourier Anal. Appl. **29**, 46 (2023)

Colombeau Type Extensions, Assymptotic Scales

Dimitris Scarpalezos

Abstract Here, we present the basic ideas behind Colombeau-type algebras and present the notion of "asymptotic scales" to construct such algebras. We present the case of ultradifferentiables and conditions of comparison of regularities between generalized functions and ultradistributions with which they are strongly associated.

Keywords Ultradiferentiable · Generalised functions · Comparison of regularities · Extension of operators in algebras

1 Introduction

When Dirac introduced the "Dirac function" δ he wrote: Let δ be a function on $\mathbb{R}$ that takes the value zero outside zero, becomes infinite at zero, and has integral 1. Of course, this has no mathematical meaning, and Schwartz noticed that it was used as an element of the dual of the test functions, and so gave a distributional meaning to this definition. Another way of understanding this definition is in the following way: "Consider a net (θ_ε) of smooth functions that when ε goes to zero its values outside zero converge to zero and at zero converges to $+\infty$. Moreover, its integral converges to one." This point of view was the source of the construction of various types of Colombeau differential algebras in which the duals of functions spaces we are interested in are embedded into (see, e.g. [1, 2, 9]).

Those algebras appear as factor spaces of "moderate nets" by the space of "negligible nets", while the meaning of a moderate and a negligible net depends on the dual space we want to embed. This embedding is obtained by convolution with a specific type of "Dirac-type" net, and so we will first present the general definition of "asymptotic scales" that will be used to give a meaning of "moderateness" and of "negligibility" in each case. These notions are closely related to the notion of very weak solution introduced by Garetto and Ruzhansky in [6].

D. Scarpalezos (✉)
University of Paris, Paris, France
e-mail: dim.scarpa@gmail.com

M. Chatzakou et al. (eds.), *Direct and Inverse Problems with Applications*,
Trends in Mathematics 12, https://doi.org/10.1007/978-3-031-98645-1_10

We will then show how the "projective" constructions of such algebras have, in a natural way, an ultrametric-type Hausdorff topology, which allows one to give meaning to the well-posedness of the problem one considers. We also recall some interesting properties of such constructions and see how these ideas apply to the Colombeau algebra introduced in [2]. Finally, we will present some results concerning the comparison of regularities in this case.

2 Colombeau Algebras and Colombeau Extensions Case

The Colombeau algebras were introduced to be able to face nonlinear equations with distribution data or linear equations with highly non-regular coefficients. In Colombeau algebras, one can embed distributions in a way that respects both derivation and products of smooth functions. However, L. Schwartz proved that it is impossible to embed distributions in an algebra, preserving both derivatives and products of continuous functions.

Definition 1 Let E be a Fréchet space, with topology induced by the countable family of seminorms (μ_n), $n \in \mathbb{N}$. We consider the family of moderate nets

$$\mathcal{E}_M[E] = \{(g_\varepsilon) \in E^{(0,1)} \text{ s.t. }, \forall n \in \mathbb{N} \exists a \in \mathbb{R}\, \exists C_n > 0 \text{ s.t. } \mu_n(g_\varepsilon) < C_n \varepsilon^a\}$$

and the family of negligible nets

$$\mathcal{N}[E] = \{(h_\varepsilon) \in E^{(0,1)} \text{ s.t. }, \forall n \in \mathbb{N} \forall a \in \mathbb{R}\, \exists C_n > 0 \text{ s.t. } \mu_n(g_\varepsilon) < C_n \varepsilon^a\}.$$

The factor space

$$\mathcal{G}[E] = \mathcal{E}_M[E]/\mathcal{N}[E]$$

is called the Colombeau extension of the space E. If (g_ε) is a moderate net, its equivalence class will be noted by $[(g_\varepsilon)]$.

In this Colombeau extension, we can establish an ultrametric type topology in the following way, see e.g. [5, 7, 11, 12].

Definition 2 Let $g = [(g_\varepsilon)] \in \mathcal{G}[E]$. For any $n \in \mathbb{N}$ we consider the "n-valuation" of g, defined by:

$$v_n(g) = \sup\{a \in \mathbb{R} \text{ s.t. } \mu_n(g_\varepsilon) = o(\varepsilon^a)\}.$$

For $g = [(g_\varepsilon)] \in \mathcal{G}[E]$ we also define the following family of ultrametric-type pseudonorms $|||g|||_n = e^{v_n(g)}$. It is easy to verify that with these pseudonorms, the space $\mathcal{G}[E]$ has a Hausdorff ultrametric type topology. This topology is called the "sharp topology" of the Colombeau extension of the space.

If $E = \mathbb{R}$, then $\mathcal{G}[E]$ is the ring of generalized real constants $\widetilde{\mathbb{R}}$ (the same can be considered for complex numbers). If $E = \mathbb{R}^n$, we have the space of generalized points. For Ω an open subset of $\mathbb{R}^d$, we consider a sequence of open sets Ω_n such that

$$\Omega_n \subset\subset \Omega_{n+1}$$

and their union is the whole open set Ω, i.e. we have $\bigcup_n \Omega_n = \Omega$. We usually equip the space $C^\infty(\Omega)$ with the topology induced by the countable family of seminorms $P_{n,m}$ defined as being the supremum of the absolute values of all derivatives up to order m on the set Ω_n. The topology of the Colombeau extension of $C^\infty(\Omega)$ described in the above way is called the Colombeau algebra of generalised functions on Ω and is denoted by $\mathcal{G}(\Omega)$. One can easily verify that it is a differential topological algebra on the algebraic structure of generalised constants for the respective sharp topologies.

This ring is not a field, but it contains a quite large field obtained by considering the generalised constants that admit an asymptotic expansion of some possibly negative terms of the form $a_n\varepsilon^{m_n}$, where the powers (m_n) are strictly increasing to infinity. Thus, our Colombeau extensions can also be seen as topological differential modules over a topological field, see e.g. [11]. One can also verify that if L is a continuous linear map between two metrisable, by countable families of seminorms spaces, then it extends as a continuous linear map between the respective Colombeau extensions, i.e., our constructions are in some sense "functorial". This implies that if we have a continuous linear partial differential equation and the solution operator is also linear continuous, then the equation is "well-posed" even in the frame of Colombeau extensions.

The usual way to define an embedding of $\mathcal{D}'(\Omega)$ in $\mathcal{G}(\Omega)$ is the following: consider a smooth function θ such that $\widehat{\theta} \in \mathcal{D}(\mathbb{R}^d)$ and is equal to one on a neighbourhood of zero. This function has as integral one, and all its higher moments are zero. As we have partition of unity, we can suppose without loss of generality that we only have to establish the embedding for distributions with compact support in Ω Consider the net of mollifiers

$$\theta_\varepsilon(\cdot) = \frac{1}{\varepsilon^d}\theta(\frac{\cdot}{\varepsilon}).$$

By convoluting the distribution T by this net of mollifiers we obtain a moderate net of smooth functions and the class of the restriction of this net on Ω will be the image of this distribution with compact support in our algebra. Analogous ways of emmbeddings will be developed in other Colombeau type algebras for the embedding of ultradistributions.

3 Asymptotic Scales and Extensions Based on Them

For the general definition of "asymptotic scale" we refer to [3, 4].

Definition 3 A family $\mathcal{A}$ of continuous positive functions on $(0, 1)$ decreasing strictly to zero as ε decreases to zero parametrized by a directed set $\mathcal{I}$ directed by a relation noted as the usual inequality $<$ will be said to constitute an asymptotic family if

(1) $\forall (m,l) \in \mathcal{I} \times \mathcal{I}$ s.t. $m > i$, $a_m(\varepsilon) = o(a_l(\varepsilon))$,
(2) $\forall (m,l) \in \mathcal{I} \times \mathcal{I}\, \exists b \in \mathcal{I}$ s.t. $a_b(\varepsilon) = o(a_m(\varepsilon)a_l(\varepsilon))$.

The first scale used is the scale (ε^k), $k \in \mathbb{N}$. Given such a scale, we can define the $\mathcal{A}-$ Colombeau of a Fréchet space E as previously (keeping the same notations in the following way.

Definition 4 A net of elements of E topologized by a family of seminorms (μ_n) is said to be $\mathcal{A}-$moderate if it belongs to

$$\mathcal{E}_M^{\mathcal{A}}[E] = \{(g_\varepsilon) \in E^{(0,1)} \text{ s.t. }, \forall n \in \mathbb{N}, \exists a \in \mathcal{A}, \exists C_n > 0 \text{ s.t.} \mu_n(g_\varepsilon) < C_n(a(\varepsilon^{-1}))\},$$

and it will be called $\mathcal{A}-$negligible if it belongs to

$$\mathcal{N}[E] = \{(h_\varepsilon) \in E^{(0,1)} \text{ s.t. }, \forall n \in \mathbb{N}, \forall a \in \mathcal{A}, \exists C_n > 0 \text{ s.t. } \mu_n(g_\varepsilon) < C_n(a(\varepsilon))\}.$$

As previously the factor space of "moderates" by "negligibles" will be the $\mathcal{A}-$ Colombeau extension of E and will be noted $\mathcal{G}^{\mathcal{A}}[E]$. When E is the set of smooth functions on an open set Ω of $\mathbb{R}^d$, we write $\mathcal{G}^{\mathcal{A}}(\Omega)$. Here, we can also define a "sharp topology" by defining a basis of neighbourhoods of zero.

If $a \in \mathcal{A}$, we define

$$B(n,a) = \{g = [(g_\varepsilon)] \text{ s.t. } \mu_n(g_\varepsilon) = o(a(\varepsilon))\}.$$

In this case, we also have generalized constants defined after the given scale, and we easily conclude that those extensions are topological modules over the ring of corresponding generalized constants for their sharp topologies. In the same way, as in the usual Colombeau case, all continuous linear or bilinear mappings can extend as continuous linear or bilinear mappings of the corresponding $\mathcal{A}$-Colombeau extensions . Thus, we have the fact that if we "generalize" a topological differential algebra, we obtain a topological differential algebra of Colombeau type on the corresponding ring of generalized constants.

The concept of asymptotic scale also allows another way of making a "generalization" (which will be used in the case of a generalisation of Roumieu regular functions); for simplicity we will present only the case of generalized constants.

Definition 5 A net (r_ε) will be said "inductively moderate" for our scale if

$$\forall a \in \mathcal{A}, r_\varepsilon = o(a^{-1}(\varepsilon)),$$

and will be said "inductively negligible" if

$$\exists a \in \mathcal{A} \text{ s.t. } r_\varepsilon = o(a(\varepsilon)).$$

Their factor space will be the ring of "inductively generalized constants" for that scale.

This kind of generalisations will be used for the ultradiferentiable Roumieu case [2], but it will not be presented it in this work.

4 Generalized Beurling Ultradifferentiable Functions

In this work will only be presented the Beurling case and the presentation of "weight function" [2]. Let us now recall the definition of the kind of weight function we will consider.

Definition 6 A positive function ω defined on the set of non-negative reals, will be said to be a "weight function" good for defining a non quasi-analytic ultradifferentiable function space if it satisfies the following properties:

(1) $\omega(0) = 1$,
(2) $\forall(a, b), \omega(a + b) \leq \omega(a) + \omega(b)$,
(3) $\lim_{t\to\infty} \frac{\omega(t)}{log(1+t)} = +\infty$,
(4) $\int_{\mathbb{R}} \frac{\omega(t)}{1+t^2} dt < \infty$.

Let us fix such a weight function ω. We will say that a test function ϕ defined on an open subspaceΩ of $\mathbb{R}^d$ is a Beurling ultradiferentiable test function if it has compact support and for any $a > 0$ and for any compact K containing its support $||\phi||_{K,l,1}$ is bounded where this semi-norm is defined by

$$||\phi||_{K,l,1} = \int_{\mathbb{R}} e^{l\omega(\xi)} |\widehat{\phi}(\xi)| d\xi.$$

The space of such test functions will be noted $\mathcal{D}^{(\omega)}(\Omega)$ (analogous globally equivalent definitions can be given with L_2 norms or supnorms, see [2], which define the same topology). The dual of this space will be called the space of Beurling ultradistributions (since we only work with the same weight function ω we will not remind it all the time). A smooth function f will be said to be Beurling ultradifferentiable for ω if for any such test function ϕ the product ϕf also belongs to this test function space. The seminorm $||\phi f||_{K,l,1}$ will be defined as the infimum of all $||\phi||_{K,l,1}$ where ϕ is such a test function which takes the value 1 on K. The space of such functions will be noted $\mathcal{E}^{(\omega)}$.

The "generalization" of this space is called the space of (ω)-Beurling ultradifferentiable generalized functions and it will be the Colombeau-type extension of this space based on the asymptotic scale

$$\mathcal{A}^{\omega} = \{e^{-k\omega(\frac{1}{\varepsilon})}, k \in \mathbb{R}^+\},$$

and will be noted

$$\mathcal{G}^{(\omega)}(\Omega).$$

The spaces of corresponding "moderate" and "negligible" nets will be noted respectively $\mathcal{E}_M^{(\omega)}(\Omega)$ and $\mathcal{N}^{(\omega)}(\Omega)$, and their factor space will be noted $\mathcal{G}^{(\omega)}(\Omega)$. The embedding of ultradistributions in our algebra is obtained as usually by considering an adequate net (θ_ε), and for any ultradistribution T with compact support and we consider the class of the net obtained by convoluting T with this net of mollifiers (see [2]).

As in the general case of generalizations with the help of an asymptotic scale, we can establish a sharp topology based on the asymptotic scale we just defined, see [8]. A basis of neighborhoods of zero is of the form $\{B_{k,l,m}\}$ where

$$B_{k,l,m} = \{g = [(g_\varepsilon)] \text{ s.t. } \|g_\varepsilon\|_{k,l,1} = o(\varepsilon^{-l\omega(\frac{1}{\varepsilon})})\}.$$

5 Regularities and Comparison of Regularities

Both in the frame of the usual Colombeau theory as well as in Colombeau Beurling's ultradifferentiable frame, notions of regularity of generalized functions have been introduced. In the usual Colombeau theory by M. Oberguggenberger in many published or unpublished papers (the notion was already clear before the nineties and in Beurling ultradifferentiable case in the thesis of Andreas Denrouwere and in [2]. This notion allows microlocal analysis in those frames, and it was proved that a distribution (resp. ultra distribution of Beurling type) is regular if and only if its image in the respective generalized algebra is regular. Here, we will remind old results of a more general theorem of comparison between the regularity of a distribution and a regular generalized function which "approaches" it in some precise sense (see [11–13]). Let us remind the notions of regularity in those frames.

Definition 7 A Colombeau generalized function $g = [(g_\varepsilon)]$ is said to be regular if for any compact subset K there exists a positive real m such that for any multi-index a the supremum on K of derivatives of order a is $o(\frac{1}{\varepsilon^m})$.

In Colombeau's theory, there are notions of "association" of generalized functions.

Definition 8 A distribution T and a generalized function $g = [(g_\varepsilon)] \in \mathcal{G}(\Omega)$ are said to be (simply) associated if

$$\forall \rho \in \mathcal{D}(\Omega), \langle T - g_\varepsilon, \rho\rangle \to 0.$$

They are said to be strongly associated if for all compact subset K of Ω there exists a positive real b such that for any test function ρ with support included in K we have:

$$\langle T - g_\varepsilon, \rho\rangle = o(\varepsilon^b).$$

We say that they are strictly associated (or equal in the distribution sense) if for any test function ρ we have:

$$[(\langle T - g_\varepsilon, \rho \rangle)] = O.$$

This is the case between a compact distribution T and its image $[(T * \theta_\varepsilon)]$ and it is easy to verify that $[(\langle T - T * \theta_\varepsilon, \rho \rangle)] = 0$.

The following theorem was established (see [11, 13]).

Theorem 1 *If a regular Colombeau generalized function g is strongly associated with a distribution T, then T is smooth.*

The proof is obtained first by supposing without loss of generality that all terms considered have compact support and then, by translating in Fourier transform settings both the notions of strong association and the notion of regularity and then by choosing, for every ξ large enough an adequate ε. We thus prove that the Fourier transform of T and all its derivatives are rapidly decreasing and thus that T is smooth. This can be useful when we consider an equation as the limit of a parametrized family of equations, and we want to know, if the net of solutions converges to some distribution and, by this theorem, conclude under some conditions, results about the regularity of this distribution. In the case of Beurling ultradiferentiable generalized functions, we have the following definition of regularity.

Definition 9 a Beurling generalized function $g = [(g_\varepsilon)] \in \mathcal{G}^{(\omega)}(\Omega)$ is said to be regular

$$g = [(g_\varepsilon)] \in \mathcal{G}^{(\omega),\infty}(\Omega),$$

if for any compact subset K there exists $m > 0$ such that

$$\forall l > 0; \, ||g_\varepsilon||_{K,l,1} = o(e^{m\omega(\frac{1}{\varepsilon})}).$$

The notions of simple and strict associations are the same, and the notion of strong association is analogous to the usual Colombeau algebra case.

Definition 10 A Beurling ultradistribution T is said to be strongly associated to $g = [(g_\varepsilon)] \in \mathcal{G}^{(\omega)}(\Omega)$ if for any compact subset K of Ω there exists a positive real b such that for any Beurling test function ρ with support in K we have

$$|\langle T - g_\varepsilon, \rho \rangle| = o(e^{-b\omega(\frac{1}{\varepsilon})}).$$

We have the following theorem.

Theorem 2 *If a Beurling generalized regular ultra differentiable function g is strongly associated with a Beurling ultra distribution T, then T is a Beurling ultra differentiable function.*

The proof here also passes first by supposing without loss of generality that all terms considered have support in some compact set K, and then we use the translation of the hypothesis and the conclusion in the Fourier transform frame and the adequate choice of ε for $|\xi|$ large enough. The proof can be found in [8].

6 Equalities

Finally, a result which seems simple (but its proof is not) is the following (it is valid as well in the Beurling ultradifferentiable case).

Theorem 3 *If a generalized function is invariant by translation, it is a generalized constant.*

The proof is rather tricky and uses Baire theorem, see [10]. In all Colombeau-type algebras it is known that if $h = [(h_\varepsilon)]$ is a generalized function such that for any test function ρ of the corresponding type, we have

$$[(\langle h_\varepsilon, \rho\rangle)] = 0.$$

This does not imply that h is zero (see for example, the difference between two different mollifier nets used for the embedding of distributions in the algebra of generalized functions). However, in [10], the following theorem was proven.

Theorem 4 *If there exists $k \in \mathbb{N}$ such that for every $\rho \in C^k(\Omega)$ with compact support we have:*

$$[(\langle h_\varepsilon, \rho\rangle)] = 0$$

then

$$h = [(h_\varepsilon)] = 0.$$

Moreover, the following theorem also holds.

Theorem 5 *If $h = [(h_\varepsilon)]$ is a regular generalized function and for all test functions ρ we have*

$$[(< h_\varepsilon, \rho >)] = 0$$

then $h = 0$.

Analogous theorems for $\mathcal{G}^{(\omega)}$ also hold, see [8]. In this case their proofs are easier because we can use norms based on L_2, and there is a convenient L_2 criterion for negligibility.

Acknowledgements I would like to thank Prof. Michael Ruzhansky and Marianna Chatzakou for their invitation to Ghent University and for giving this lecture.

References

1. Colombeau, F.: Elementary Introduction to New Generalized Functions. North Holland Publishing Co., Amsterdam (1985)
2. Debrouwere, A., Vindas, J., Vernaeve, H.: Optimal embedding of of ultradistributions into differential algebras. Monatsh. Math. **186**, 407–438 (2018)
3. Delcroix, A., Scarpalezos, D.: Asymptotic scales asymptotic algebras, integral transforms. Spec. Funct. **6**, 181–190 (1998)
4. Delacroix, A., Scarpalezos, D.: Sharp topologies asymptotic algebras. Monatsh. Math. **129** (2000)
5. Garetto, C.: Pseudo-differential Operators with generalized symbols and regularity theory, Ph.D. thesis, University of Turin (2004)
6. Garetto, C., Ruzhansky, M.: Hyperbolic second order equations with non-regular time dependent coefficients. Arch. Ration. Mech. Anal. **217**, 113–154 (2015)
7. Garetto, C.: Pseudodifferential operators in algebras of generalized functions and global hypoellipticity. Acta Appl. Math. **80**, 123–174 (2004)
8. Khelif, A., Scarpalezos, D.: Bjork-Debrouwere Algebras, Topologies and Comparison of Regularities. In preparation
9. Oberguggenberger, M.: Multiplication of distributions and application to partial differential equations. In: Pitman Research Notes in Mathematics Series, vol. 259, Longman, Harlow (1992)
10. Piliovic, S., Scarpalezos, D., Valmorin, V.: Equalities in algebras of generalized functions. Forum Math. **18**, 789–801 (2006)
11. Scarpalezos, D.: Colombeau generalized functions, Topological structures; Microlocal properties a simplified pint of view prepublications mathematiques de Paris 7: CNRS, URA212 (1993)
12. Scarpalezos, D.: Colombeau generalized functions : generalized functions, topological structures , microlocal properties. A simplified point of view, Part I. Bull. Cl. Sci. Math. Nat. Sci. Math. **25**, 89–114 (2000)
13. Scarpalezos, D.: Colombeau generalized functions : generalized functions, topological structures , microlocal properties. A simplified point of view, Part II. Publ. Inst. Math. (Beograd. (N.S.)) **76**, 111–125 (2004)

Analytical Solutions to the Laplace Equation on a Hemispherical Domain

Cristiana Sebu, Andrei Amaira, and Michael Pidcock

Abstract In this paper, we derive analytical solutions to the Laplace equation in a hemispherical domain subject to two different idealized Neumann boundary conditions. The solutions are given as infinite series, and their convergence is analysed. The theoretical results have been validated by comparing them with numerical results obtained using EIDORS.

Keywords Laplace equation · Neumann boundary conditions · Direct problems · Electrical impedance tomography

MSC2020 35C10 · 35C15 · 35J05 · 35J08 · 35J25

1 Introduction

Electrical Impedance Tomography (EIT) is a non-invasive, portable, low-cost technology developed to image the electrical properties of the material within an object from electrical boundary measurements [1]. EIT has been actively investigated as a medical imaging modality, including breast cancer detection [2].

Several EI mammographic systems have been developed over the years [3–6] and, more recently, fully portable and autonomous systems embedded in a brassiere [7–9] have also been proposed.

The inverse problem of EIT is to reconstruct the unknown conductivity σ distribution inside a simply connected domain $\Omega \in \mathbb{R}^n$, $n = 2$ or 3, with boundary $\partial\Omega$, from the knowledge of the Neumann-to-Dirichlet map $\mathcal{R}_\sigma$ defined by

C. Sebu (✉) · A. Amaira
Department of Mathematics, University of Malta, Msida, Malta
e-mail: cristiana.sebu@um.edu.mt

M. Pidcock
School of Engineering Computing and Mathematics, Oxford Brookes University, Oxford, UK

M. Chatzakou et al. (eds.), *Direct and Inverse Problems with Applications*, Trends in Mathematics 12, https://doi.org/10.1007/978-3-031-98645-1_11

$$\mathcal{R}_\sigma : \widetilde{H}^{-\frac{1}{2}}(\partial\Omega) \to H^{\frac{1}{2}}(\partial\Omega),$$
$$j = \sigma \frac{\partial \Phi}{\partial n}\bigg|_{\partial\Omega} \mapsto \Phi|_{\partial\Omega},$$

where

$$\nabla \cdot (\sigma \nabla \Phi) = 0 \quad \text{in } \Omega\,, \tag{1}$$

and $\widetilde{H}^{-\frac{1}{2}}$ space consists of $H^{-\frac{1}{2}}$ functions with mean zero, j is the applied current density and Φ is the induced potential.

Reconstruction methods for the full nonlinear inverse problem often rely on iterative optimization techniques, which, though accurate, can be computationally intensive and slow, especially for three-dimensional problems [2]. To address this problem, alternative direct algorithms to reduce computational demands have been developed, for example, by using linearized integral equations [6, 10, 11]. These approaches require the knowledge of $\mathcal{R}_1$, i.e. the solution to equation (1) subject to the same applied boundary currents j for a constant conductivity distribution $\sigma = 1$. This solution is often referred to as the associated potential function. In practice, however, the current density is not applied continuously along the boundary but rather through a number of electrodes attached to the boundary, and the number of measurements is finite. Therefore, not only is a full knowledge of $\mathcal{R}_\sigma$ not available in practice, but also deriving analytical expressions for the associated potential functions, even in the continuum model, could be impossible except in the case of specific boundary domain geometries (circle, sphere, hemisphere etc.).

In this paper, we derive the potential functions associated with a hemispherical domain subject to two different Neumann boundary conditions corresponding to standard trigonometric current patterns applied to two idealised electrode models, the Dirac and the Heaviside models. This work is motivated by the further development of a linearized integral equation reconstruction method similar to [6, 10, 11] for the proposed three-dimensional hemispherical brassiere design for EI mammography in [8, 9].

2 Associated Potential Functions in a Hemispherical Domain

In this paper, the domain of interest, Ω is the upper hemisphere of radius R, namely

$$\Omega = \left\{ \mathbf{x} = (\rho, \theta, \varphi) \in \mathbb{R}^3 : 0 \le \rho \le R, 0 \le \theta \le \frac{\pi}{2}, 0 \le \varphi < 2\pi \right\}.$$

By the method of images, the Neumann Green's function for the interior of Ω [12], denoted by $\mathcal{H}_N(\mathbf{P}, \mathbf{Q})$, is given by

$$\mathcal{H}_N(\mathbf{P}, \mathbf{Q}) = \mathcal{G}_N(\mathbf{P}, \mathbf{Q}) + \mathcal{G}_N(\mathbf{P}, \tilde{\mathbf{Q}}).$$

$\mathbf{P} = \mathbf{P}(\rho', \theta', \varphi')$ and $\mathbf{Q} = \mathbf{Q}(\rho, \theta, \varphi)$ are arbitrary points in Ω, $\widetilde{\mathbf{Q}} = \widetilde{\mathbf{Q}}(\rho, \pi - \theta, \varphi)$ is the reflection of $\mathbf{Q}$ in the xy-plane and $\mathcal{G}_N(\mathbf{P}, \mathbf{Q})$ is the Neumann Green's function for the interior of a sphere of radius R [13]:

$$\mathcal{G}_N(\mathbf{P}, \mathbf{Q}) = \frac{1}{4\pi}\left(\frac{1}{r_{\mathbf{PQ}}} + \frac{R}{\rho' r_{\mathbf{P}\overline{\mathbf{Q}}}} - \frac{1}{\rho'} - \frac{1}{R}\right) + \frac{1}{R}\sum_{n=1}^{\infty}\sum_{m=-n}^{n}\frac{1}{n(2n+1)}\left(\frac{\rho\rho'}{R^2}\right)^n Y_n^{m*}(\theta, \varphi) Y_n^m(\theta', \varphi'),$$

where $r_{\mathbf{PQ}}$ denotes the distance between points $\mathbf{P}$ and $\mathbf{Q}$ and likewise, $r_{\mathbf{P}\overline{\mathbf{Q}}}$ denotes the distance between points $\mathbf{P}$ and $\overline{\mathbf{Q}} = \overline{\mathbf{Q}}(R^2/\rho, \theta, \phi)$, as shown in Fig. 1, and $Y_n^m(\theta, \varphi)$ denotes the spherical harmonic of degree n and order m whilst $Y_n^{m*}(\theta, \varphi)$ denotes its complex conjugate [14].

If $\mathbf{P} = \left(R, \theta', \varphi'\right) \in \partial\Omega$, the expression for the Green's function $\mathcal{G}_N(\mathbf{P}, \mathbf{Q})$ can be simplified to [8]:

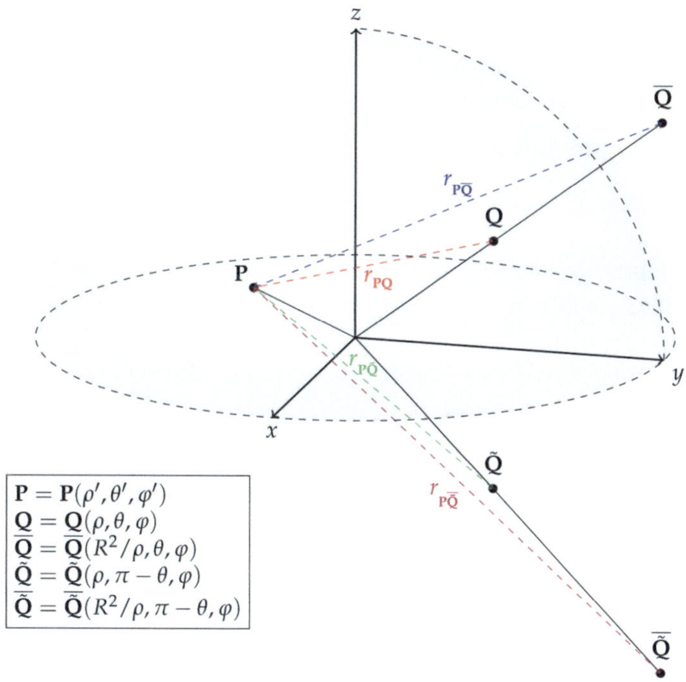

Fig. 1 Hemisphere domain with points of interest

$$\mathcal{G}_N(\mathbf{P},\mathbf{Q}) = -\frac{1}{2\pi R} + \frac{1}{R}\sum_{n=1}^{\infty}\sum_{m=-n}^{n}\frac{1}{n}\left(\frac{\rho}{R}\right)^n Y_n^{m*}(\theta,\varphi)Y_n^m(\theta',\varphi').$$

Note that the Neumann Green's function is symmetric [13], i.e. $\mathcal{G}_N(\mathbf{P},\mathbf{Q}) = \mathcal{G}_N(\mathbf{Q},\mathbf{P})$.

The associated potential function, $\Phi_0^{(\alpha)}$, the solution to the following boundary value problem:

$$\Delta\Phi_0^{(\alpha)}(\mathbf{x}) = 0, \quad \mathbf{x} = (\rho,\theta,\varphi) \in \Omega, \tag{2}$$

$$\frac{\partial\Phi_0^{(\alpha)}}{\partial n}(\mathbf{x}) = j^{(\alpha)}(\mathbf{x}), \quad \mathbf{x} = (R,\theta,\varphi) \in \partial\Omega,$$

$$\frac{\partial\Phi_0^{(\alpha)}}{\partial n}(\mathbf{x}) = 0, \quad \mathbf{x} = (\rho,\frac{\pi}{2},\varphi) \in \partial\Omega, \tag{3}$$

such that $\int_{\partial\Omega}\Phi_0^{(\alpha)}(\mathbf{x})\,d\mathbf{x} = 0$, can be obtained applying Green's second identity

$$\Phi_0^{(\alpha)}(\mathbf{x}) = \int_{\partial\Omega}\mathcal{H}_N(\mathbf{x},\mathbf{y})j^{(\alpha)}(\mathbf{y})\,d\mathbf{y}\,,$$

which in spherical coordinates reads

$$\Phi_0^{(\alpha)}(\rho,\theta,\varphi) = R^2\int_0^{\pi/2}d\theta'\sin(\theta')\int_0^{2\pi}d\varphi'\mathcal{H}_N(\rho,\theta,\varphi;R,\theta',\varphi')j^{(\alpha)}(\theta',\varphi')\,. \tag{4}$$

Analytical expressions for $\Phi_0^{(\alpha)}$ are presented next for two different idealised Neumann boundary conditions $j^{(\alpha)}$, the Dirac and the Heaviside electrode models.

2.1 Dirac Electrode Model

In this section, trigonometric current patterns of the following form are considered.

$$j^{(\alpha)}(\theta',\varphi') = \begin{cases}\cos(\alpha\varphi')\delta(\theta'-\theta_0), & \text{for } \alpha = \overline{1,\ 5},\\ \sin((\alpha-5)\varphi')\delta(\theta'-\theta_0), & \text{for } \alpha = \overline{6,\ 10},\end{cases} \tag{5}$$

for some fixed $0 < \theta_0 < \pi/2$. This Neumann boundary condition corresponds to trigonometric current patterns applied to a ring electrode of infinitesimal width placed at $\theta' = \theta_0$, as shown in Fig. 2. Thus, for $\alpha = \overline{1,\ 5}$, Eq. (4) becomes

Fig. 2 Ring electrode for the Dirac model

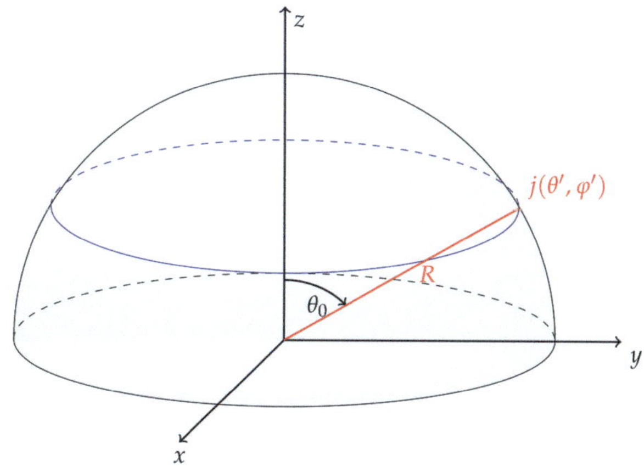

$$\Phi_0^{(\alpha)}(\rho,\theta,\varphi) = R^2 \sin(\theta_0) \int_0^{2\pi} d\varphi' \mathcal{H}_N(\rho,\theta,\varphi; R,\theta_0,\varphi') \cos(\alpha\varphi')$$
$$= R \sin(\theta_0) \cos(\alpha\varphi) \times$$
$$\times \sum_{m=\alpha}^{\infty} \frac{4m-2\alpha+1}{2m-\alpha} \left(\frac{\rho}{R}\right)^{2m-\alpha} \frac{(2m-2\alpha)!}{(2m)!} P_{2m-\alpha}^{\alpha}(\cos\theta_0) P_{2m-\alpha}^{\alpha}(\cos\theta), \quad (6)$$

where $P_{2m-\alpha}^{\alpha}$ are associated Legendre functions [14]. Similarly, for $\alpha = \overline{6,\ 10}$ and $\beta = \alpha - 5$,

$$\Phi_0^{(\alpha)}(\rho,\theta,\varphi) = R^2 \sin(\theta_0) \int_0^{2\pi} d\varphi' \mathcal{H}_N(\rho,\theta,\varphi; R,\theta',\varphi') \sin\big((\alpha-5)\varphi'\big)$$
$$= R \sin(\theta_0) \sin(\beta\varphi) \times$$
$$\times \sum_{m=\beta}^{\infty} \frac{4m-2\beta+1}{2m-\beta} \left(\frac{\rho}{R}\right)^{2m-\beta} \frac{(2m-2\beta)!}{(2m)!} P_{2m-\beta}^{\beta}(\cos\theta) P_{2m-\beta}^{\beta}(\cos\theta_0). \quad (7)$$

The series in (6) and (7) can be shown to be convergent for any $\theta = \theta_0$ and $\rho < R$. However, the series is divergent when $\rho = R$ [8].

2.2 *Heaviside Electrode Model*

Since the series expressions of the associated potential functions for a Dirac electrode model in (6) and (7) are not convergent on the boundary $\partial\Omega$, the boundary conditions in Eq. (5) are replaced here by more realistic Neumann boundary conditions whereby trigonometric current patterns are injected on a ring electrode of finite width ϵ centred

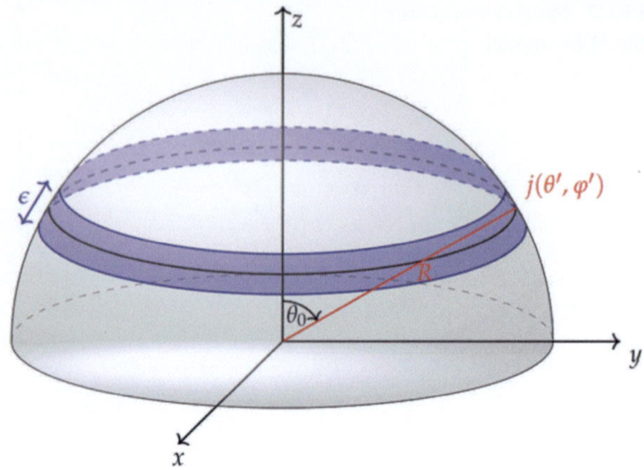

Fig. 3 Ring electrode for Heaviside model

at $\theta = \theta_0$, such that

$$j^{(\alpha)}(\theta', \varphi') = \begin{cases} \frac{1}{\epsilon}\cos(\alpha\varphi'), & \text{for } \alpha = \overline{1,\,5},\ \theta_0 - \frac{\epsilon}{2R} \le \theta' \le \theta_0 + \frac{\epsilon}{2R} \\ \frac{1}{\epsilon}\sin((\alpha-5)\varphi'), & \text{for } \alpha = \overline{6,\,10},\ \theta_0 - \frac{\epsilon}{2R} \le \theta' \le \theta_0 + \frac{\epsilon}{2R} \\ 0, & \text{for } 0 \le \theta' < \theta_0 - \frac{\epsilon}{2R} \text{ and } \theta_0 + \frac{\epsilon}{2R} < \theta' \le \frac{\pi}{2}, \end{cases} \tag{8}$$

as shown in Fig. 3. In this case, the expression for $\Phi_0^{(\alpha)}$ in (4) becomes

$$\Phi_0^{(\alpha)}(\rho, \theta, \varphi) = R^2 \int_{\theta_0 - \frac{\epsilon}{2R}}^{\theta_0 + \frac{\epsilon}{2R}} d\theta' \sin(\theta') \int_0^{2\pi} d\varphi' \mathcal{H}_N(\rho, \theta, \varphi; R, \theta', \varphi') j^{(\alpha)}(\varphi').$$

Hence, the expression for the associated potential is given by

$$\Phi_0^{(\alpha)}(\rho, \theta, \varphi) = \frac{R}{\epsilon}\cos(\alpha\varphi) \sum_{m=\alpha}^{\infty} \frac{4m - 2\alpha + 1}{4m - 2\alpha} \frac{(2m - 2\alpha)!}{(2m)!} \left(\frac{\rho}{R}\right)^{2m-\alpha} P_{2m-\alpha}^{\alpha}(\cos\theta) \times$$
$$\times \int_{\theta_0 - \frac{\epsilon}{2R}}^{\theta_0 + \frac{\epsilon}{2R}} d\theta' \sin(\theta') P_{2m-\alpha}^{\alpha}(\cos\theta'), \quad \text{for} \quad \alpha = \overline{1,\,5}, \tag{9}$$

and similarly,

$$\Phi_0^{(\alpha)}(\rho, \theta, \varphi) = \frac{R}{\epsilon}\sin(\beta\varphi) \sum_{m=\alpha}^{\infty} \frac{4m - 2\beta + 1}{4m - 2\beta} \frac{(2m - 2\beta)!}{(2m)!} \left(\frac{\rho}{R}\right)^{2m-\beta} P_{2m-\beta}^{\beta}(\cos\theta) \times$$
$$\times \int_{\theta_0 - \frac{\epsilon}{2R}}^{\theta_0 + \frac{\epsilon}{2R}} d\theta' \sin(\theta') P_{2m-\beta}^{\beta}(\cos\theta'), \quad \text{for} \quad \beta = \alpha - 5 = \overline{1,\,5}. \tag{10}$$

It can be shown that the infinite series expressions of the associated potential functions in Eqs. (9) and (10) are convergent everywhere in Ω [8]. Moreover, the

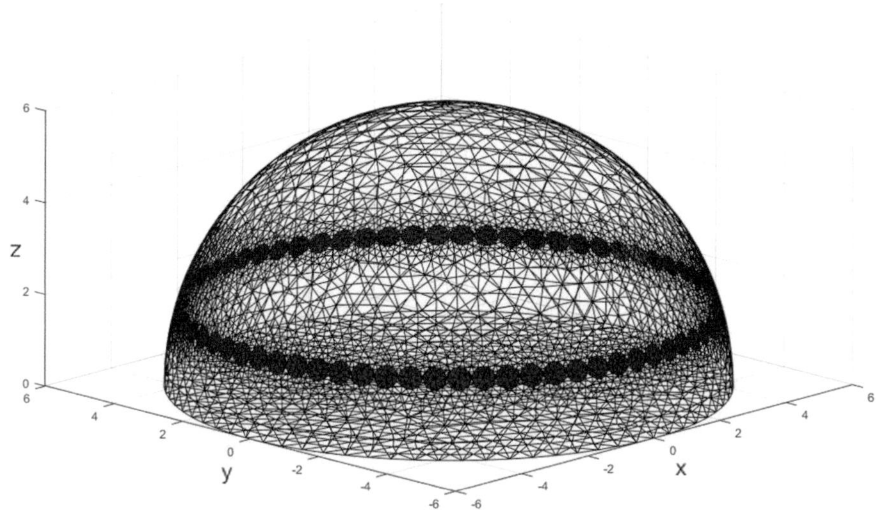

Fig. 4 Finite element mesh of a hemisphere of radius $R = 6$ cm with a ring of 70 electrodes of radius 0.25 cm centered at $\theta_0 = \arcsin\left(\frac{5.75}{6}\right)$

theoretical results also agree for all $\alpha = \overline{1,\ 10}$ in Eq. (8) with the simulated potentials obtained by solving numerically the forward problem (2)–(3) in EIDORS [15]. To get a close discrete approximation of our idealised ring electrode in the Heaviside model, we considered a ring of 70 electrodes of radius $r_e = 0.25$cm centered at $\theta_0 = \arcsin\left(\frac{5.75}{6}\right)$ on a hemisphere of radius $R = 6$ cm, as shown in Fig. 4, to which trigonometric current patterns of the form

$$I^{(\alpha)}(\varphi_i) = \begin{cases} \cos(\alpha\varphi_i), & \text{for } \alpha = \overline{1,\ 5}, \\ \sin((\alpha - 5)\varphi_i), & \text{for } \alpha = \overline{6,\ 10}, \end{cases} \tag{11}$$

where φ_i is the azimuthal angle of the centre of an electrode, $i = \overline{1,\ 70}$, were applied.

Figures 5 illustrates the plot of the associated potential on the boundary given by Equation (9) for an electrode of width equal to $\varepsilon = 2r_e = 0.5$ cm, $\alpha = 1$, $\rho = R$ and $\theta = \theta_0$ in Eq. (8) as a function of the polar angle $\varphi \in [0,\ 2\pi)$ where the integral over θ' was approximated numerically by using Gauss-Legendre quadrature in Matlab, together with the plot of the simulated voltage in EIDORS that corresponds to $\alpha = 1$ in Eq. (11).

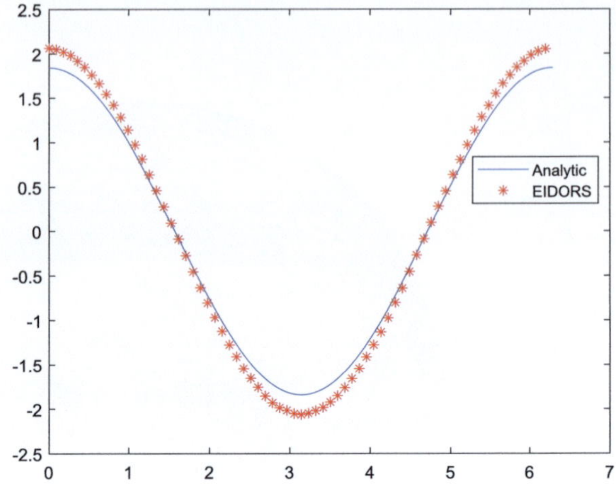

Fig. 5 Plot of the associated potential functions for $\alpha = 1$ on the boundary, $\Phi_0^{(1)}(R, \theta_0, \varphi), 0 \leq \phi < 2\pi$

3 Conclusion

In this paper, we presented the analytical expressions of the solutions to the Laplace equation on a hemispherical domain subject to two continuous idealised Neumann boundary conditions. The solution in the case of the more realistic Heaviside electrode model was convergent everywhere in the considered domain, including the boundary, and was validated by comparison with the numerical solution obtained in EIDORS. This solution will be a very useful tool in the development of a three-dimensional linearized integral equation imaging algorithm for EI mammography.

Acknowledgements This work has been partially supported by the University of Malta Research Excellence Fund Programme under the grant number I21LU01. The authors would also like to thank to Professor Andy Adler, Carleton University, Canada, for his assistance with the numerical implementations in EIDORS.

References

1. Borcea, L.: Electrical impedance tomography. Inverse Prob. **18**(6), R99–R136 (2002)
2. Holder, D.S.: Electrical Impedance Tomography: Methods, History and Applications. CRC Press (2004)
3. Ye, G., Lim, K.H., George Jr, R.T., Ybarra, G.A., Joines, W.T., Liu, Q.H: 3D EIT for breast cancer imaging: System, measurements, and reconstruction. Microwave Opt. Technol. Lett. **50** (12), 3261–3271 (2008)
4. Halter, R.J., Hartov, A., Paulsen, K.D.: A broadband high-frequency electrical impedance tomography system for breast imaging. IEEE Trans. Biomed. Eng. **55**(2), 650–659 (2008)
5. Sze, G.: Detection of breast cancer with electrical impedance mammography. University of Sussex, Thesis (2012)

6. Georgi, K.H., Hähnlein, C., Schilcher, K., Sebu, C., Spiesberger, H.: Conductivity reconstructions using real data from a new planar electrical impedance tomography device. Inverse Prob. Sci. Eng. **21**(5), 801–822 (2013)
7. Hong, S., Lee, K., Ha, U., Kim, H., Lee, Y., Kim, Y., Yoo, H.J.: A 4.9 mΩ-sensitivity mobile electrical impedance tomography IC for early breast-cancer detection system. IEEE J. Solid-State Circuits **50**(1), 245–257 (2015)
8. Amaira, A.: Reconstruction Algorithms for Electrical Impedance Mammography. University of Malta (2023)
9. Sebu, C., Amaira, A.: Three-dimensional brassiere design for electrical impedance tomography and numerical conductivity reconstructions in EIDORS. In: Proceedings of the 11th International Conference "Inverse Problems: Modelling and Simulation" (IPMS 2024), p. 6. Malta. Accepted for publication to Springer (2024)
10. Hähnlein, C., Schilcher, K., Sebu, C., Spiesberger, H.: Conductivity imaging with interior potential measurements. Inverse Prob. Sci. Eng. **19**(5), 729–750 (2011)
11. Sebu, C., Amaira, A., Curmi, J.: A linearized integral equation reconstruction method of admittivity distributions using electrical impedance tomography. Eng. Anal. Boundary Elem. **150**, 103–110 (2023)
12. Langford, J.: Neumann comparison theorems in elliptic pdes. Potential Anal. **43**(3), 415–459 (2015)
13. Dassios, G., Sten, J.C.E.: On the neumann function and the method of images in spherical and ellipsoidal geometry. Math. Methods Appl. Sci. **35**(4), 482–496 (2012)
14. Gradshteyn, I.S., Ryzhik, I.M.: Table of Integrals, Series, and Products. Academic, New York (2014)
15. Adler, A., Grychtol, B.: EIDORS version 3.12. In: 24th International Conference on Biomedical Applications of Electrical Impedance Tomography (EIT 2024), Hangzhou, China (2024)

Uniform Spectral Asymptotics for High-Contrast Periodic Media

Ilia V. Kamotski, Valery P. Smyshlyaev, and Shane Cooper

Abstract For an elliptic operator with high-contrast periodic matrix-inclusion coefficients in $\mathbb{R}^n$, $n \geq 2$, we obtain improved results on uniform approximations of typical Floquet-Bloch eigenvalues in terms of those of an explicit two-scale limit operator. As a result, we obtain not only improved rates for convergence of the spectra to the limit spectrum, that contains band gaps, but also improved uniform estimates for an explicit asymptotics for the integrated density of states.

Keywords Floquet-bloch spectrum · High-contrast · Two-scale operator approximations

MSC2020 35B40 · 35B27

1 Formulation and Background

We consider a self-adjoint elliptic operator $\mathcal{A}_\tau = -\nabla \cdot A_\tau \nabla$ in $\mathbf{R}^n$, $n \geq 2$, depending on a positive parameter of contrast τ as follows. The coefficients $A_\tau(x)$ are assumed to be periodic with periodicity unit cell $\square = [-1/2, 1/2]^n$, i.e. $A_\tau(x + m) = A_\tau(x)$, $\forall x \in \mathbb{R}^n$ and $\forall m \in \mathbb{Z}^n$, and to take two different values in the isolated periodic inclusions B and in their complements:

$$A_\tau(x) = \begin{cases} \tau, & x \in \square \backslash B, \\ 1, & x \in B. \end{cases} \tag{1}$$

I. V. Kamotski · V. P. Smyshlyaev (✉) · S. Cooper
Department of Mathematics, University College London, London, UK
e-mail: v.smyshlyaev@ucl.ac.uk

I. V. Kamotski
e-mail: i.kamotski@ucl.ac.uk

S. Cooper
e-mail: s.cooper@ucl.ac.uk

M. Chatzakou et al. (eds.), *Direct and Inverse Problems with Applications*, Trends in Mathematics 12, https://doi.org/10.1007/978-3-031-98645-1_12

The boundary of inclusion B is assumed to be Lipschitz, $\overline{B} \subset \left(-\frac{1}{2}, \frac{1}{2}\right)^n$ for simplicity, and the "matrix" (periodically extended $\square\backslash B$) is therefore connected. The interest is in the behaviour of operator $\mathcal{A}_\tau$ and its spectral characteristics for large contrasts τ.

According to Floquet-Bloch theory, the spectrum of $\mathcal{A}_\tau$ has a band structure:

$$\mathrm{Sp}\,\mathcal{A}_\tau \;=\; \bigcup_{m=1}^{\infty} \Big[\; \min_{\theta\in\square^*} \lambda_\tau^{(m)}(\theta)\,,\; \max_{\theta\in\square^*} \lambda_\tau^{(m)}(\theta) \Big], \tag{2}$$

where $\square^* = [-\pi, \pi]^n$ is the dual periodicity cell, and the "dispersion relations"

$$0 \;\le\; \lambda_\tau^{(1)}(\theta) \le \lambda_\tau^{(2)}(\theta) \le \cdots \le \lambda_\tau^{(m)}(\theta) \le \cdots\,, \quad \lambda_\tau^{(m)}(\theta) \to +\infty \;\text{ as } m \to \infty,$$

are the eigenvalues of the (non-negative self-adjoint) operators in Hilbert space $L^2(\square)$, $\mathcal{A}_\tau(\theta) = -(\nabla + i\theta)\cdot A_\tau(\nabla + i\theta)$ with the periodicity conditions on $\square$.

It was shown in [1] that, as $\tau \to +\infty$, $\mathrm{Sp}\,\mathcal{A}_\tau$ converges, in the sense of Hausdorff (with unknown rate), to a limit set displaying band gaps. (So gaps do open for large enough τ in $\mathrm{Sp}\,\mathcal{A}_\tau$.) Zhikov [2], for a spectrally equivalent operator $\mathcal{B}_\varepsilon = -\,\varepsilon^2\nabla\cdot A_{\varepsilon^{-2}}(\cdot/\varepsilon)\nabla$ with ε-periodic coefficients where $\varepsilon = \tau^{-1/2}$, showed that the limit set is the spectrum of a non-negative self-adjoint two-scale limit operator $\mathcal{B}_0$ defined on a "bigger" (two-scale) Hilbert space $\mathbb{H}_0 \subset L^2\,(\mathbb{R}^n \times \square)$. He also showed that $\mathcal{B}_\varepsilon$ converges to $\mathcal{B}_0$ in the sense of (strong) "two-scale (pseudo-) resolvent convergence". In [3, 4] certain estimates on asymptotics of integrated density of states

$$m_\tau(\lambda) \;:=\; (2\pi)^{-n} \sum_{m=1}^{\infty} \mathrm{meas}\,\left\{\theta \in \square^* \,:\, \lambda_\tau^{(m)}(\theta) \;\le\; \lambda\right\}, \tag{3}$$

as $\tau \to \infty$ were obtained, demonstrating a spectral concentration near the right ends of generic limit bands (that are simple eigenvalues of the Dirichlet Laplacian in B with a non-zero mean value for the corresponding eigenfunctions). Robust estimates on the rate of convergence of the spectra near the above typical eigenvalues, explicit in terms of the inclusions' shapes and geometry, were obtained in [5] via decomposition of quasi-periodic and "electrostatic" solution operators with a subsequent analysis of related resonances by tools of layer potential theory.

In our recent work [6], we proposed a general approach for uniform two-scale type operator approximations for a broad class of spectral problems and applied it, in particular, to operators akin to $\mathcal{B}_\varepsilon$, equivalently $\mathcal{A}_\tau$. Resulting resolvent approximations and operator estimates in terms of the two-scale limit operator led us, in particular, to the following approximations and estimates for $\lambda_{\varepsilon,\theta}^{(k)} := \lambda_{\varepsilon^{-2}}^{(k)}(\theta)$:

$$\left|\lambda_{\varepsilon,\theta}^{(k)} \;-\; \Lambda_{\theta/\varepsilon}^{(k)}\right| \;\le\; C\,\varepsilon, \quad \forall\theta \in \square^*, \quad \forall 0 < \varepsilon < 1, \tag{4}$$

for a constant C that is independent of ε and θ (although dependent on k).

In (4), the explicit approximations $\Lambda_\xi^{(k)}$, $\xi \in \mathbb{R}^n$, $k \in \mathbb{N}$, are as follow. They are eigenvalues of operators $\mathbb{L}_\xi$ entering a direct fiber-integral decomposition of the Zhikov's two-scale limit operator $\mathcal{B}_0$. Namely, for $\mathbb{L} = \mathcal{F}^{-1}\mathcal{B}_0\mathcal{F}$ where $\mathcal{F}$ is Fourier transform in the "macroscopic" variable

$$\mathbb{L} \;=\; \int_{\mathbb{R}^n}^{\oplus} \mathbb{L}_\xi \,\mathrm{d}\xi .$$

For every $\xi \in \mathbb{R}^n$, $\mathbb{L}_\xi$ is a nonnegative self-adjoint operator in a (complex) Hilbert space $\mathcal{H}_0 = \mathbb{C} \dotplus L^2(B)$, with $L^2(B)$ understood here as the subspace of functions from $L^2(\square)$ which vanish outside B. (So $\mathcal{H}_0$ consists of functions from $L^2(\square)$ which are constant in $\square\backslash B$.) The spectrum of $\mathbb{L}_\xi$ is discrete with eigenvalues $0 \le \Lambda_\xi^{(1)} \le \Lambda_\xi^{(2)} \le \ldots \Lambda_\xi^{(m)} \le \ldots$, $\Lambda_\xi^{(m)} \to +\infty$ as $m \to \infty$, whose associated eigenfunctions $\psi_\xi^{(m)} = c_m + v_m \not\equiv 0$ are such that $c_m \in \mathbb{C}$ and $v_m \in H^2(B) \cap H_0^1(B)$ solve

$$\begin{cases} \left(A^{\mathrm{h}}\,\xi \cdot \xi\right) c_m \;=\; \Lambda_\xi^{(m)} \left(c_m \;+\; \displaystyle\int_B v_m(x)\,\mathrm{d}x \right), \\ -\,\Delta\, v_m(x) \;=\; \Lambda_\xi^{(m)} \Big(c_m \;+\; v_m(x) \Big), \qquad x \in B. \end{cases} \tag{5}$$

In (5), A^h is the "perforated domain" homogenized matrix, which is a positive definite symmetric $n \times n$ matrix with associated quadratic form

$$A^{\mathrm{h}}\,\xi \cdot \xi \;=\; \min_{u \in H^1_{\mathrm{per}}(\square)} \int_{\square\backslash B} |\xi \,+\, \nabla u(x)|^2 \mathrm{d}x,$$

where $H^1_{\mathrm{per}}(\square)$ denotes $\square$-periodic functions from $H^1_{\mathrm{loc}}(\mathbb{R}^n)$.

It follows from (5) that if $\Lambda \in \mathrm{Sp}\,\mathbb{L}_\xi$, then either Λ is an eigenvalue of the Dirichlet Laplacian in B ($\Lambda \in \mathrm{Sp}\,(-\Delta_D)$) which has an associated eigenfunction ϕ with a zero mean over B ($\langle\phi\rangle := \int_B \phi(x)\mathrm{d}x = 0$), or $\Lambda \notin \mathrm{Sp}\,(-\Delta_D)$ and $\beta(\Lambda) = A^h\xi \cdot \xi$. Here $\beta(\Lambda)$ is the Zhikov's β-function [2, 7] well-defined on $\mathbb{R}\backslash\mathrm{Sp}\,(-\Delta_D)$:

$$\beta(\Lambda) \;:=\; \Lambda \,+\, \Lambda^2 \int_B b(x\,,\,\Lambda)\,\mathrm{d}x,$$

where $b(\cdot\,,\Lambda) = (-\Delta_D \,-\, \Lambda)^{-1}\,\mathbf{e}$ ($\mathbf{e}$ denotes the unity function) i.e. is the unique solution of

$$-\,\Delta b \;-\; \Lambda\, b \;\equiv\; 1 \quad \text{in } B, \quad b = 0 \text{ on } \partial B.$$

Function $\beta(\Lambda)$ has spectral decomposition

$$\beta(\Lambda) \;=\; \Lambda \,+\, \Lambda^2 \sum_{m=1}^{\infty} \frac{|\langle\phi_m\rangle|^2}{\Lambda_m \,-\, \Lambda}, \tag{6}$$

where ϕ_m are L^2-orthonormal eigenfunctions of $-\Delta_D$ corresponding to $\Lambda_m \in \mathrm{Sp}\,(-\Delta_D)$. Remark that, for $B = B_a$ a ball of radius $a < 1/2$ centred at the origin, $\beta(\Lambda)$ is found explicitly in terms of trigonometric or Bessel functions. In particular, for $n = 3$,

$$\beta(\Lambda) \;=\; \Lambda\left(1 - 4\pi a^3/3\right) \,+\, 4\pi a\left(1 - a\Lambda^{1/2}\mathrm{cotan}\left(\Lambda^{1/2}a\right)\right),$$

see e.g. [8, p. 419]. Generally, $\beta(\Lambda)$ vanishes on $\mathrm{Sp}\,\mathbb{L}_0\backslash\mathrm{Sp}(-\Delta_D)$, strictly increases and blows-up near eigenvalues Λ_m with $\langle\phi_m\rangle \neq 0$, see (6).

The Hausdorff limit of the spectrum of $\mathcal{B}_\varepsilon$, equivalently (2) of $\mathcal{A}_\tau$, coincides with

$$\begin{aligned}\mathrm{Sp}\,\mathcal{B}_0 \;=\; \mathrm{Sp}\,\mathbb{L} \;&=\; \overline{\bigcup_{\xi\in\mathbb{R}^n} \mathrm{Sp}\,\mathbb{L}_\xi} \\ &=\; \bigcup_{m=1}^{\infty}\,\left[\,\mu_m\,,\,\Lambda_m\,\right] \;=\; \overline{\left\{\lambda \notin \mathrm{Sp}\,(-\Delta_D)\;:\;\beta(\lambda)\geq 0\right\}} \,\cup\, \mathrm{Sp}\,(-\Delta_D),\end{aligned}$$

where $\mu_m := \Lambda_0^{(m)} \leq \Lambda_m$, eigenvalues of "electrostatic problem", are $\Lambda_\xi^{(m)}$ for $\xi = 0$. We have shown in [6] that

$$\mathrm{dist}_{[0,b]}\Big(\,\mathrm{Sp}\,\mathcal{B}_\varepsilon\,,\;\mathrm{Sp}\,\mathcal{B}_0\Big) \;\leq\; C(b)\varepsilon, \quad \forall b \geq 0\,, \quad \forall 0 < \varepsilon < 1. \tag{7}$$

In (7), $\mathrm{dist}_{[a,b]}(X, Y) := \max\big(\mathrm{dist}([a, b] \cap X, Y), \mathrm{dist}([a, b] \cap Y, X)\big)$ where $\mathrm{dist}(X, Y) := \sup_{x\in X} \mathrm{dist}(x, Y)$ is the (non-symmetric) distance, with the adopted convention that $\mathrm{dist}(\emptyset, A) = \mathrm{dist}(A, \emptyset) = 0$ for any set A.

The above uniform spectral approximations and estimates (4) appear to also provide us with, in particular, some new results on the asymptotics of the integrated density of states, (3), see Corollary 7.17 of [6]. The estimates (4) were obtained by us in [6], however, as a direct specialization of a general approach, developed therein, to the simple model (1) with the high-contrast real scalar coefficients. An advantage of our general approach, and of the results based on it, is that they apply with little or no change for more general models. In particular, the uniform estimate (4) and some of its corollaries apply for general measurable complex Hermitian matrix-valued $A_\tau(x) = \tau A_1(x) + A_2(y)$ where A_1 and A_2 are supported in $\square\backslash B$ and $\overline{B}$ respectively and possess uniform boundedness and uniform positive definiteness properties (and in fact even more generally for systems, cf. recent work [9] and further references therein). However, specifically for (1) (and in fact more generally for *real*-valued symmetric matrices A_1 and A_2) our results in [6] can be strengthened further as follows.

2 New Results for Model (1)

The following strengthening of estimate (4) holds.

Theorem 1 *Let $k = 1$, or $k > 1$ and the eigenvalue Λ_k of the Dirichlet Laplacian $-\Delta_D$ on B is simple with the associated eigenfunction ϕ_k having a non-zero mean over B: $\langle\phi_k\rangle := \int_B \phi_k(x)\mathrm{d}x \neq 0$. Let additionally, for $k > 1$, the preceding (not necessarily simple) Dirichlet eigenvalue $\Lambda := \max_{j<k} \Lambda_j$ have at least one eigenfunction ϕ such that $\langle\phi\rangle \neq 0$. Then*

$$\left|\lambda^{(k)}_{\varepsilon,\theta} \; - \; \Lambda^{(k)}_{\theta/\varepsilon}\right| \;\leq\; C_k\,\varepsilon^2, \qquad \forall\theta \in \square^*, \qquad \forall\, 0 < \varepsilon < 1, \tag{8}$$

with a constant C_k independent of ε and θ (although dependent on k).

Notice that, compared to (4), the new estimate (8) displays the improved rate ε^2 although under additional assumptions on the Dirichlet eigenvalue Λ_k. These assumptions however appear to hold generically in a certain sense, see e.g. a discussion in Remark 3.13 of [1] and further references in [3].

Theorem 1 leads us in turn to the following strengthening, compared to [6] Corollary 7.17, of the estimates on the asymptotics of the integrated density of states $m_\tau(\lambda)$, see (3), (where $\tau = \varepsilon^{-2}$) inside a non-degenerate limit band-gap $[\mu_k, \Lambda_k]$:

Corollary 1 *Under the assumptions of Theorem 1, let $\mu_k := \Lambda^{(k)}_0$, i.e. $\Lambda^{(k)}_\xi$ for $\xi = 0$. Then, $\mu_k < \Lambda_k$ i.e. the limit band $[\mu_k, \Lambda_k]$ is of a positive length, and $\beta(\lambda) > 0$ on (μ_k, Λ_k). Let C_k be the constant in the inequality (8). Then for $\mu_k + 2C_k\tau^{-1} < \lambda < \Lambda_k - 2C_k\tau^{-1}$ one has*

$$m_\tau(\lambda) \;=\; k \,-\, 1 \,+\, (2\pi)^{-n}\,\omega_n\tau^{\,-n/2}\frac{[\beta(\lambda)]^{n/2}}{\left(\det A^{\mathrm{h}}\right)^{1/2}}\left\{1 \;+\; R_\tau(\lambda)\right\}, \tag{9}$$

where ω_n is the measure of the unit ball in $\mathbb{R}^n$, and

$$|R_\tau(\lambda)| \;\leq\; D_k\,\frac{\tau^{-1}}{(\lambda-\mu_k)\,(\Lambda_k-\lambda)} \quad \textit{with a constant } D_k \textit{ independent of } \tau \geq 1 \textit{ and } \lambda. \tag{10}$$

Proofs of Theorem 1 and Corollary 1, as well as some other related results, will be given elsewhere.

Notice that (9)–(10) provides a uniform asymptotic approximation to $m_\tau(\lambda)$ as long as $\Lambda_k - \lambda \gg \tau^{-1}$ and $\lambda - \mu_k \gg \tau^{-1}$, i.e. with exception of small neighbourhoods of both ends of the limit bands $[\mu_k, \Lambda_k]$. It follows from (6) and other properties of $\beta(\lambda)$ that $\beta(\mu_k) = 0$, $\beta'(\mu_k) > 0$, $\beta(\lambda)$ strictly increases on $[\mu_k, \Lambda_k)$, and

$$\beta(\lambda) \;=\; \Lambda_k^2\frac{|\langle\phi_k\rangle|^2}{\Lambda_k \,-\, \Lambda} \;+\; O(1), \quad \text{as} \quad \Lambda \to \Lambda_k.$$

This makes (9)–(10) for $\Lambda_k - \lambda \gtrsim \tau^{-1}$ improving [3, 4] demonstrating concentration of the spectral density near $\lambda = \Lambda_k$. When $\lambda - \mu_k = O\left(\tau^{-1}\right)$, the above leads to certain upper and lower bounds for $m_\tau(\lambda)$. The related analysis also leads to certain estimates on the location of the left and right ends of the kth band of the original operator $\mathcal{A}_\tau$, equivalently $\mathcal{B}_\varepsilon$, for large τ, cf. [5].

References

1. Hempel, R., Lienau, K.: Spectral properties of periodic media in the large coupling limit. Commun. Partial Differ. Equ. **25**(7–8), 1445–1470 (2000)
2. Zhikov, V.V.: On spectrum gaps of some divergent elliptic operators with periodic coefficients. St. Petersburg Math. J. **16**(5), 773–790 (2005)
3. Friedlander, L.: On the density of states of periodic media in the large coupling limit. Commun. Partial Differ. Equ. **27**(1–2), 355–380 (2002)
4. Selden, J.: Periodic operators in high-contrast media and the integrated density of states function. Commun. Partial Differ. Equ. **30**, 1021–1037 (2005)
5. Lipton, R., Viator, R.: Creating band gaps in periodic media. SIAM Multiscale Model. Simul. **15**(4), 1612–1650 (2017)
6. Cooper, S., Kamotski, I.V., Smyshlyaev, V.P.: Quantitative multiscale operator-type approximations for asymptotically degenerating spectral problems (2025). arXiv:2307.13151
7. Zhikov, V.V.: On an extension of the method of two-scale convergence and its applications. Sb. Math. **191**(7), 973–1014 (2000)
8. Babych, N.O., Kamotski, I.V., Smyshlyaev, V.P.: Homogenization of spectral problems in bounded domains with doubly high contrasts. Netw. Heterogen. Media **3**(3), 413–436 (2008)
9. Cherednichenko, K., Kiselev, A., Velčić, I., Žubrinić, J.: Effective behaviour of critical-contrast PDEs: micro-resonances, frequency conversion, and time dispersive properties. II . Comm. Math. Phys. **406**, 72 (2025)

Asymptotic Mean Value Formulas for the p-Laplacian in the Euclidean Space and in the Heisenberg Group

Bianca Stroffolini

Abstract The classical mean value property is a fundamental characterization of harmonic functions and illustrates the deep connection between the Laplacian and Brownian motion. Building on this interplay between probability and analysis, Peres, Schramm, Sheffield, and Wilson introduced in their influential paper (Journal of the AMS, 22(1), 167–210, 2009) a probabilistic game known as Tug-of-War, which provides a game-theoretic framework for approximating solutions to the infinity-Laplacian. This framework has since been generalized in various directions to handle a wider class of nonlinear PDEs, notably the p-Laplacian. A key tool in this context is the asymptotic mean value property, which serves as a nonlinear analogue of the classical mean value formula and forms the basis of a semidiscrete dynamic programming principle at scale ε. The solutions to this principle are shown to converge to the unique viscosity solution of the corresponding Dirichlet problem. The scope of this course is to present an overview of: the classical mean value formulas for the Laplacian; the asymptotic mean value properties for the infinity-Laplacian and the p-Laplacian and their connections to viscosity solutions.

Keywords p-Laplacian · Natural p-means · Dirichlet problem · Discrete approximations · Asymptotic mean value properties · Convergence · Generalized viscosity solutions · Heisenberg group

MSC2020 35J92 · 35D40 · 49L20 · 49L25 · 35R02 · 35B05 · 35J62

1 Linear Equations

Several characterizations of harmonic functions involving surface and solid averages on Euclidean balls are given in the literature. More generally, using the classical potential theory and estimates on the level sets of the fundamental solution, the following characterization holds for subharmonic functions:

B. Stroffolini (✉)
Department of Mathematics, University Federico II, Naples, Italy
e-mail: bstroffo@unina.it

M. Chatzakou et al. (eds.), *Direct and Inverse Problems with Applications*, Trends in Mathematics 12, https://doi.org/10.1007/978-3-031-98645-1_13

Theorem 1 *Let $\Omega \subset \mathbb{R}^n$ be an open set and let $u : \Omega \mapsto [-\infty, \infty)$ be an upper semicontinuous function (u.s.c.) such that the set $S(u) := \{x \in \Omega : u(x) > -\infty\}$ contains at least one point of every component of Ω. Then the following are equivalent:*

- *u verifies the comparison with respect to harmonic functions on every subdomain $V \subset \bar{V} \subset \Omega$;*
- *for every $x \in \Omega$ and $B_\epsilon(x) \subset \Omega$, we have*

$$u(x) \leq \mathbf{S}_\epsilon u(x) := \fint_{\partial B(x,\epsilon)} u(y) d\mathcal{H}^{n-1}(y),$$

where

$$\fint_{\partial B(x,\epsilon)} u(y) d\mathcal{H}^{n-1}(y) = \frac{1}{\mathcal{H}^{n-1}(\partial B(x,\epsilon))} \int_{\partial B(x,\epsilon)} u(y) d\mathcal{H}^{n-1}(y);$$

- *for every $x \in \Omega$ and $B_\epsilon(x) \subset \Omega$, we have*

$$u(x) \leq \mathbf{A}_\epsilon u(x) := \fint_{B(x,\epsilon)} u(y) dy = \frac{1}{|B(x,\epsilon)|} \int_{B(x,\epsilon)} u(y) dy;$$

- *(Blaschke) for every $x \in S(u)$, we have*

$$(\limsup_{\epsilon \to 0}) \frac{1}{\epsilon^2} \Big[\fint_{\partial B(x,\epsilon)} u(y) d\mathcal{H}^{n-1}(y) - u(x) \Big] \geq 0;$$

- *(Privaloff) for every $x \in S(u)$, we have*

$$(\limsup_{\epsilon \to 0}) \frac{1}{\epsilon^2} \Big[\fint_{B(x,\epsilon)} u(y) dy - u(x) \Big] \geq 0;$$

- *for every $x \in \Omega$ the function $r \mapsto \mathbf{S}_r u(x)$ is increasing and $\lim_{r \to 0} \mathbf{S}_r u(x) = u(x)$;*
- *for every $x \in \Omega$ the function $r \mapsto \mathbf{A}_r u(x)$ is increasing and $\lim_{r \to 0} \mathbf{A}_r u(x) = u(x)$.*

Similarly, one can characterize superharmonic functions. Sharp versions of Blaschke and Privaloff conditions were proved by Sachs.

Based on the Taylor formula, if u is smooth then it holds that

$$\fint_{B_\epsilon(x)} u(y) dy = u(x) + \frac{1}{2(n+2)} \mathrm{trace}(D^2 u) \epsilon^2 + o(\epsilon^2), \epsilon \to 0$$

for any ball $B(x,\epsilon) \subset \Omega$, ϵ small. Otherwise, the validity of the asymptotic mean value formula (AMVF)

$$u(x) = \fint_{B(x,\epsilon)} u(y)dy + o(\epsilon^2)$$

for all $x \in \Omega$, ϵ small enough and any ball $B(x,\epsilon) \subset \Omega$ is equivalent to being the solution of the Laplace equation.

Even more, one can use a semidiscrete approximation. The Dynamic Programming Principle (DPP) at scale ε built upon the mean value formula is:

$$u_\varepsilon(x) = \fint_{B(x,\epsilon)} u(y)dy. \tag{1}$$

Suppose that $g\colon \partial\Omega \to \mathbb{R}$ is continuous. In order to consider (1) for all points $x \in \overline{\Omega}$ and for all $0 < \varepsilon \le 1$, one has to extend the function g continuously to a function G in the ε-boundary strip

$$\Gamma_\varepsilon = \{x \in \mathbb{R}^n \setminus \Omega : \operatorname{dist}(x,\Omega) \le \varepsilon\},$$

and consider the problem

$$\begin{cases} u_\varepsilon(x) = \fint_{B_\varepsilon(x)} u_\varepsilon(y)\,dy, & x \in \Omega, \\ u_\varepsilon(x) = \quad G(x), & x \in \Gamma_\varepsilon. \end{cases}$$

One may wonder if the solutions u_ϵ may eventually converge to the unique solution u of the Dirichlet problem:

$$\begin{cases} \Delta u = 0 & \text{in } \Omega, \\ u \; = g & \text{on } \partial\Omega. \end{cases} \tag{2}$$

This is indeed the case, if g and Ω are Lipschitz. For fixed ϵ small enough, there exists a unique solution u_ϵ and

$$u_\epsilon \to u \quad \text{uniformly in } \bar{\Omega}.$$

Furthermore, the mean value formula (1) can be replaced by a discrete one:

$$u_\epsilon(x) = \frac{1}{n}\sum_{i=1}^{n}\{\frac{1}{2}u(x+\epsilon e_i) + \frac{1}{2}u(x-\epsilon e_i)\}.$$

where $(e_1,\ldots,e_n)$ is the canonical basis. Also in this case, one may consider the associated (DPP) and prove the uniform convergence to the solution of (2).

If we replace the Laplace equation $\Delta u = 0$ by a linear elliptic equation with constant coefficients $Lu = \sum_{i,j} a_{ij}u_{x_i x_j} = 0$, then mean value formulas hold for appropriate ellipsoids instead of balls. Bonfiglioli and Lanconelli in [7] proved extensions of Blaschke and Privaloff Theorems to the subelliptic case.

Another classical viewpoint occurs via the Brownian motion. Let $(\mathbb{B}_\tau^x)_{\tau\geq 0}$ be a standard Brownian motion starting at $x \in \Omega$. Denote by $\tau_{\partial\Omega}$ the exit time:

$$\tau_{\partial\Omega} = \inf\{\tau > 0 : \mathbb{B}_\tau^x \notin \Omega\} > 0.$$

Note that $\mathbb{B}_{\tau_{\partial\Omega}}^x \in \partial\Omega$.

Theorem 2 (Kakutani 1944) *u is harmonic in* Ω *iff for any ball* $B(x,r) \subset \Omega$, *we have*

$$u(x) = \mathbb{E}[u(\mathbb{B}_{\tau_{\partial B(x,r)}}^x)].$$

The roadmap we are going to follow in the sequel also for nonlinear equations is summarised below:

AMVF (continuous, discrete) $\Longleftrightarrow$ DPP (usually discrete) $\Longleftrightarrow$ PDE (usually continuous).

Theorem 3 *For an appropriate real function u we have a meta-equivalence among the following facts:*

1. *u satisfies a Mean Value Formula (in an appropriate asymptotic sense),*
2. *u can be approximated by solutions to a Dynamic Programming Principle associated to a game or control problem,*
3. *u solves a (possibly nonlinear) PDE.*

We would like to stress that the approach described above is very flexible: can be applied to Euclidean spaces, Riemannian manifolds, Sub-Riemannian manifolds (Heisenberg group), graphs (trees), metric-measure spaces.

2 The ∞-Laplacian and Tug-of-War Means

The ∞-Laplace equation was discovered by Aronsson [3]:

$$\Delta_\infty u = \langle D^2 u \nabla u, \nabla u\rangle$$

This can be viewed as the formal limit for $p \to \infty$ of the p-Laplacian equations

$$\Delta_p u \equiv |\nabla u|^{p-4}\{|\nabla u|^2 \Delta u + (p-2)\Delta_\infty u\} = 0.$$

To see the latter we divide out the quantity $(p-2)|\nabla u|^{p-4}$ and let $p \to \infty$ in the expression

$$\frac{|\nabla u|^2 \Delta u}{p-2} + \Delta_\infty u = 0.$$

The variational solution of the ∞-Laplacian can be indeed obtained as the limit of p-harmonic functions, see [5]. Given a Lipschitz continuous function g defined on the boundary of a bounded domain $\Omega \subset \mathbb{R}^n$, for $p > n$, we first minimize the variational integral

$$I(u) = \int_\Omega |\nabla u(x)|^p dx$$

among all functions in $C(\bar{\Omega}) \cap C^1(\Omega)$ with boundary values g. There exists a unique minimizer, say u_p. Using a compactness argument, based on Ascoli-Arzelà Theorem, we can conclude the following result:

Proposition 1 *There exists a subsequence* $\{u_{p_h}\}$ *and a function* $u_\infty \in C(\bar{\Omega}) \cap (g + W_0^{1,\infty}(\Omega))$ *such that* $u_{p_h} \to u_\infty$ *uniformly in* Ω *and* $\nabla u_{p_h} \rightharpoonup \nabla u_\infty$ *in each* $L^s(\Omega)$.

This limit is called a variational solution of the ∞-Laplace equation. Since

$$\lim_{p\to\infty} \left\{ \int_D |\nabla u(x)|^p dx \right\}^{\frac{1}{p}} = \text{esssup}_{x\in D} |\nabla u(x)| = \|\nabla u\|_{\infty(D)}$$

for each subdomain D we have that

$$\|\nabla u_\infty\|_{\infty(D)} \le \|\nabla v\|_{\infty(D)}$$

whenever v is Lipschitz continuous in $\bar{D}$ and $v = u_\infty$ on the boundary ∂D. This property is known as Absolutely Minimizing Lipschitz Extension (AMLE).

In principle, any variational solution could depend on the subsequence. However, this is not the case since it is a viscosity solution. Let us recall the definition of viscosity subsolution and supersolution.

Definition 1 For $1 < p \le \infty$, let Ω be a bounded open set.

- An upper semicontinuous function u is a viscosity subsolution of $\Delta_p u = 0$ in Ω if and only if for all $x_0 \in \Omega$ and $\phi \in C^2$ such that $u - \phi$ has a strict max at x_0, we have $\Delta_p \phi(x_0) \ge 0$.
- A lower semicontinuous function u is a viscosity supersolution of $\Delta_p u = 0$ in Ω if and only if for all $x_0 \in \Omega$ and $\psi \in C^2$ such that $u - \psi$ has a strict min at x_0, we have $\Delta_p \psi(x_0) \le 0$.

Jensen proved in [16] the uniqueness of viscosity solution of the ∞-Laplace equation, see also [2]. A geometric approach given in terms of comparison with cones was suggested by Crandall et al. [10].

The half-cone $C = a + b|x - x_o|$ is a viscosity solution of ∞-Laplacian when $x \ne x_o$. It plays the role of a fundamental solution. If a function satisfies the comparison with cones, then it obeys the comparison principle with respect to all viscosity super and sub-solutions.

Definition 2 We say that $u \in C(\Omega)$ obeys the comparison principle from above with cones if, for each subdomain $D \subset\subset \Omega$, the inequality

$$u(x) - b|x - x_o| \leq \max\{u(y) - b|y - x_o| y \in \partial D\}$$

holds when $x \in D$, $x_o \in \mathbb{R} \setminus D$. For the comparison principle from below with cones, the required inequality is

$$u(x) - b|x - x_o| \geq \min\{u(y) - b|y - x_o|, y \in \partial D\}.$$

The function obeys the comparison principle with cones if it does so both from above and below.

In order to derive the AMVF for the ∞-Laplacian, we can proceed formally assuming that u is smooth and $\nabla u(x) \neq 0$. By Taylor expansion:

$$\begin{aligned} u(x) - \frac{1}{2}\Big\{ \max_{B_\epsilon(x)} u + \min_{B_\epsilon(x)} u \Big\} &\approx u(x) - \frac{1}{2}\Big\{ u\Big(x + \epsilon\frac{\nabla u(x)}{|\nabla u(x)|}\Big) + u\Big(x - \epsilon\frac{\nabla u(x)}{|\nabla u(x)|}\Big)\Big\} \\ &= -\frac{\epsilon^2}{2}\Delta_\infty^N u(x) + o(\epsilon^2). \end{aligned}$$

The general theorem deals with continuous functions and viscosity solutions.

Theorem 4 *A continuous function u is a viscosity solution of the ∞-Laplacian iff it satisfies the asymptotic mean value formula in the viscosity sense. This means that the two conditions hold:*

- *if $x_o \in \Omega$ and if $\phi \in C^2(\Omega)$ touches u from below at x_o, then*

$$\phi(x_o) \geq \frac{1}{2}\Big\{ \max_{B_\epsilon(x_o)} \phi + \min_{B_\epsilon(x_o)} \phi \Big\} + o(\epsilon^2).$$

  Moreover, if it happens that $\nabla\phi(x_o) = 0$, we require that the test function satisfies $\mathbb{D}^2\phi(x_o) \leq 0$.
- *if $x_o \in \Omega$ and if $\psi \in C^2(\Omega)$ touches u from above at x_o, then*

$$\psi(x_o) \leq \frac{1}{2}\Big\{ \max_{B_\epsilon(x_o)} \psi + \min_{B_\epsilon(x_o)} \psi \Big\} + o(\epsilon^2).$$

  Moreover, if it happens that $\nabla\psi(x_o) = 0$, we require that the test function satisfies $\mathbb{D}^2\psi(x_o) \geq 0$.

Let's describe the tug-of-war game. Let $\Omega \subset \mathbb{R}^n$ be a bounded domain with boundary $\partial\Omega$. A token is placed at a point $x \in \Omega$. The boundary $\partial\Omega$ is split into two parts: a favorable exit set $C \subset \partial\Omega$ for Player I and its complement $\partial\Omega \setminus C$, favorable to Player II. The players want the token to exit through their respective favorable sets. A fair coin is tossed to decide who moves next. On winning a coin toss, a player chooses the next position of the token within a ball of radius ϵ (i.e., token moves less than ϵ units). The move is made in the direction that favors their objective. The game ends when the token reaches the boundary $\partial\Omega$. The value of the game at point $x \in \Omega$ is the probability that the token exits through C, under optimal play by both players. At small scales (as $\varepsilon \to 0$), the optimal movement direction for the players is aligned with $\pm\frac{\nabla u_\epsilon}{|\nabla u_\epsilon|}$, where u is the value function of the game. These directions maximize or minimize the function u, and are determined by local gradients. As the step size $\varepsilon \to 0$, the discrete game converges to a continuous value function $u(x)$,that is a solution of the ∞-Laplace equation.

For $1 < p < \infty$ a smooth function with non-vanishing gradient satisfies at a point $x \in \mathbb{R}^n$ the (normalized) p-harmonic equation

$$\sum_{i,j=1}^{n} \left\{\delta_{ij} + (p-2)\frac{u_{x_i}(x)u_{x_j}(x)}{|\nabla u(x)|^2}\right\} u_{x_i x_j}(x) = 0,$$

if and only if

$$u(x) = \frac{\alpha}{2}\left(\sup_{B_\varepsilon(x)} u + \inf_{B_\varepsilon(x)} u\right) + \beta \fint_{B_\varepsilon(x)} u(y)\,dy + o(\varepsilon^2), \tag{3}$$

where $\alpha = \frac{p-2}{n+p}$ and $\beta = \frac{n+2}{n+p}$.

The expansion (3) suggests the consideration of solutions u_ε to the Dynamic Programming Principle (DPP)

$$u_\varepsilon(x) = \frac{\alpha}{2}\left(\sup_{B_\varepsilon(x)} u_\varepsilon + \inf_{B_\varepsilon(x)} u_\varepsilon\right) + \beta \fint_{B_\varepsilon(x)} u_\varepsilon(y)\,dy, \tag{4}$$

where $\Omega \subset \mathbb{R}^n$ is a bounded domain, $u_\varepsilon \colon \Omega \to \mathbb{R}$ and $B_\varepsilon(x) \subset \Omega$. Suppose that $g\colon \partial\Omega \to \mathbb{R}$ is continuous. In order to consider (4) for all points $x \in \overline{\Omega}$ and for all $0 < \varepsilon \leq 1$, one has to extend the function g continuously to a function G in the ε-boundary strip

$$\Gamma_\varepsilon = \{x \in \mathbb{R}^n \setminus \Omega : \ \text{dist}(x, \Omega) \leq \varepsilon\},$$

and consider the problem

$$\begin{cases} u_\varepsilon(x) = \frac{\alpha}{2}\left(\sup_{B_\varepsilon(x)} u_\varepsilon + \inf_{B_\varepsilon(x)} u_\varepsilon\right) + \beta \fint_{B_\varepsilon(x)} u_\varepsilon(y)\,dy, & x \in \Omega, \\ u_\varepsilon(x) = G(x), & x \in \Gamma_\varepsilon. \end{cases} \tag{5}$$

When the domain Ω is Lipschitz, one can solve the Dirichlet problem (5).

Manfredi, Parviainen and Rossi, using a game theory approach, proved the following Theorem:

Theorem 5 ([22]) *The family* $\{u_\varepsilon\}_{0<\varepsilon\leq 1}$ *of functions converges uniformly in* $\overline{\Omega}$ *to* u, *the unique viscosity solution to the Dirichlet problem*

$$\begin{cases} \Delta_p^N u = \displaystyle\sum_{i,j=1}^{n} \left\{ \delta_{ij} + (p-2)\frac{u_{x_i}u_{x_j}}{|\nabla u|^2} \right\} u_{x_i x_j} = 0, & in\ \Omega \\ u = g, & in\ \partial\Omega. \end{cases} \tag{6}$$

Note that viscosity solutions of the problem (6) are also weak solutions, when the normalized p-Laplacian is replaced by the variational p-Laplacian, [17].

The nonlinear mean value expression in the right-hand side of (4) is a *tug-of-war with noise* mean, suggested by connections with probability developed in [25, 26], see also [20] for related results in the Heisenberg Group $\mathbb{H}$.

3 Natural p-Means

Ishiwata, Magnanini and Wadade [15], considered new asymptotic formulas, based on the "natural p-means". Let X be compact and endowed with a positive finite Radon measure ν. Given $u \in L^p(X, \nu)$ and $1 < p \leq \infty$ there exists a unique real value $\mu_p^X(u)$ such that

$$\|u - \mu_p^X(u)\|_p = \min_{\lambda\in\mathbb{R}} \|u - \lambda\|_p .$$

The number $\mu_p^X(u)$ is called the natural p-mean of u in (X, ν). The definition extends to the case $p = 1$ when u is continuos. In few cases cases, as $p = 1, 2$, and $p = \infty$, there is an explicit formula for $\mu_p(u)$:

$$\begin{aligned} \mu_1(u) &= \mathrm{med}(u), \\ \mu_2(u) &= ⨍_X u(y)d\nu, \text{ and} \\ \mu_\infty(u) &= \tfrac{1}{2}(\mathrm{essinf}_{y\in X} u(y) + \mathrm{esssup}_{y\in X} u(y)). \end{aligned}$$

The following properties were established in [15]:

- Continuity in the L^p-norm:

$$\left| \|u - \mu_p^X(u)\|_p - \|v - \mu_p^X(v)\|_p \right| \leq \|u - v\|_p.$$

- Monotonicity : if $u \leq v$ a.e. on $L^p(X)$, then we have $\mu_p^X(u) \leq \mu_p^X(v)$.

- Affine invariance: $\mu_p^X(u+c) = c + \mu_p^X(u)$ and $\mu_p^X(\alpha\, u) = \alpha\, \mu_p^X(u)$ for $c, \alpha \in \mathbb{R}$.

Assume next that X is a subset of $\mathbb{R}^n$. The means at scale ϵ are defined as follows:

$$\mu_p(u,\epsilon)(x) = \mu_p^{\overline{B_\epsilon(x)}}(u) = \mu_p(u_\epsilon^x, 1)(0),$$

where $u_\epsilon^x(z) = u(x + \epsilon\, z)$. The function $x \mapsto \mu_p(u,\epsilon)(x)$ is continuous.

Theorem 6 ([15]) *Let $\phi \in C^2$ such that $\nabla\phi(x) \neq 0$. Then it holds:*

$$\mu_p(\phi,\epsilon)(x) = \phi(x) + \frac{\epsilon^2}{2(n+p)}\Delta_p^H\phi(x) + o(\epsilon^2)$$

as $\epsilon \to 0$, where $\Delta_p^H\phi(x)$ is the homogeneous p-Laplacian given by

$$\Delta_p^H = |\nabla u|^{2-p} div|\nabla u|^{p-2}\nabla u) = \sum_{i,j=1}^{n}\{\delta_{ij} + (p-2)\frac{u_{x_i}u_{x_j}}{|\nabla u|^2}\}u_{x_ix_j}$$

Corollary 1 ([15]) *A continuous function u is p-harmonic if and only if*

$$\mu_p(u,\epsilon)(x) = u(x) + o(\epsilon^2)$$

as $\epsilon \to 0$ in the viscosity sense.

4 DPP and Convergence in $\mathbb{R}^n$, a PDE Viewpoint

In this section, we will review a remarkable result obtained by del Teso, Manfredi, and Parviainen in [11].

For a broad class of averaging operators—those for which an asymptotic mean value formula for the p-Laplacian an be defined—they proved that the solutions of the associated dynamic programming principle at scale ϵ converge, as $\epsilon \to 0$, to the viscosity solution of the corresponding Dirichlet problem. The assumptions are general enough to include both the Tug-of-War means and more natural averaging schemes. Notably, their proof is purely PDE-based, relying on the properties of radial solutions to the p-Laplace equation within a sequence of iterated circular annuli.

Let $\Omega \subset \mathbb{R}^n$ be a bounded Lipschitz domain and $g\colon \partial\Omega \mapsto \mathbb{R}$ be a bounded Lipschitz function. As before, for $\epsilon > 0$, one consider the strip Γ_ϵ of width ϵ around the boundary of Ω. Set $\Omega_E = \Omega \cup \Gamma_1$ and let $B(\Omega)$, $B(\Omega_E)$ be the classes of bounded real functions defined on Ω and Ω_E, respectively.

Definition 3 We say that an operator $A : B(\Omega_E) \to B(\overline{\Omega})$ is an average if it satisfies the following properties:

- (Stable) $\inf_{\{y\in\Omega_E\}} \phi(y) \le A[\phi](x) \le \sup_{\{y\in\Omega_E\}} \phi(y), \forall x \in \overline{\Omega}$;
- (Monotone) If $\phi \le \psi$ in Ω_E then $A[\phi] \le A[\psi]$ in $\overline{\Omega}$;
- (Affine invariance) $A[\lambda\phi + \xi] = \lambda A[\phi] + \xi, \forall \lambda > 0, \forall \xi \in \mathbb{R}$.

Definition 4 (*AMVP*) We say that a family of averages $\{A_\epsilon\}_{\epsilon>0}$ satisfies the **(asymptotic) mean value property** for the p-Laplacian if for every $\phi \in C^\infty(\Omega_E)$ such that $\nabla\phi \ne 0$ we have

$$A_\epsilon[\phi](x) = \phi(x) + c\,\epsilon^2 \left(\Delta_p^H \phi(x)\right) + o(\epsilon^2),$$

for some constant $c > 0$ independent of ϵ and ϕ and where the constant in $o(\epsilon^2)$ can be taken uniformly for all $x \in \overline{\Omega}$.

Given $g \in C(\partial\Omega)$, let G be a continuous extension of g to Γ_1. Associated to the mean value property A_ϵ we have the dynamic programming principle at scale ϵ given by

$$\begin{cases} u_\epsilon(x) = A_\epsilon[u_\epsilon](x) & \text{in } \Omega, \\ u_\epsilon(x) = \quad G(x) & \text{on } \Gamma_\epsilon. \end{cases}$$

This is a Dirichlet problem for the fixed point of the average operator A_ϵ. We now list some properties of the family $\{A_\varepsilon\}$.

- The family of averages $\{A_\varepsilon\}$ satisfies the Uniform Stability (**US**) if for all $0 < \epsilon < 1$ there exists $u_\epsilon \in B(\Omega_E)$, a solution of (DPP_ϵ) with a bound on $\|u_\epsilon\|_{L^\infty(\Omega_E)}$ uniform in ϵ.
- The family of averages $\{A_\varepsilon\}$ satisfies the Uniform Boundedness (**UB**) if for all $0 < \epsilon < 1$ there exists $u_\epsilon \in B(\Omega_E)$, a solution of (DPP_ϵ), and we have $\inf_{\Omega_E} G \le u_\epsilon(x) \le \sup_{\Omega_E} G$ for all $x \in \Omega$.
- The family of averages $\{A_\varepsilon\}$ satisfies the Comparison Principle (**CP**) if , given u_ϵ^1 and u_ϵ^2 subsolution and supersolution of (DPP_ϵ) with boundary data G_1 and G_2 respectively, the inequality $G^1 \le G^2$ on Γ_1 implies $u_\epsilon^1 \le u_\epsilon^2$ in Ω_E.

The main convergence theorem in [11] is the following:

Theorem 7 *Let $\Omega \subset \mathbb{R}^n$ be a bounded domain and $g \in C(\partial\Omega)$. Let the family of averages $\{A_\epsilon\}_{\epsilon>0}$ satisfy the* **AMVP** *with respect to the p-Laplacian. Let also $\{u_\epsilon\}_{\epsilon>0}$ be a sequence of solutions of the corresponding (4). Then we have that*

- *if the domain Ω is of class C^2 and the family of averages $\{A_\epsilon\}_{\epsilon>0}$ satisfies* **US***, or*
- *if the domain Ω is Lipschitz and the family of averages $\{A_\epsilon\}_{\epsilon>0}$ satisfies the* **UB** *and* **CP** *properties, then*

$$u_\epsilon \to v \text{ uniformly in } \overline{\Omega} \text{ as } \epsilon \to 0,$$

where v is the unique viscosity solution of the Dirichlet problem

$$\begin{cases} -\Delta_p u = 0 & \text{in } \Omega \\ u \quad = g & \text{on } \partial\Omega. \end{cases}$$

The proof for C^2-domains follows by generalizing to the p-Laplacian a classical convergence result of Barles and Souganidis [4] who proved the convergence when the operator satisfies the strong uniqueness principle. A key role is played by the regularity properties of the Euclidean distance function.

The proof for Lipschitz domains is based on solving the problem with smooth data and use good properties of the *fundamental solution* of the p-Laplacian that allow to set up an appropriate iteration.

5 AMVF for the p-Laplacian in the Heisenberg Group $\mathbb{H}$

In this section, we would like to extend the AMVF and the convergence results in the Heisenberg Group, see [12].

The Heisenberg group $\mathbb{H} = (\mathbb{R}^3, *)$ is a Carnot Group of step two, with the group operation defined as

$$q * q' = (x, y, z) * (x', y', z') = \big(x + x', y + y', z + z' + \frac{1}{2}(xy' - yx')\big).$$

The Korányi gauge is

$$|q|_K = \big((x^2 + y^2)^2 + 16z^2\big)^{1/4}.$$

It induces a left invariant metric $d(q, q') = |q^{-1} * q'|_K$ in $\mathbb{H}$. There is a family of anisotropic dilations given by:

$$\rho_\lambda(x, y, z) = (\lambda x, \lambda y, \lambda^2 z), \qquad \lambda > 0,$$

that are group homomorphisms. The Korányi gauge and the Korányi metric are homogeneous with respect to the dilations

$$|\rho_\lambda(q)|_K = \lambda |q|_K, \; d(\rho_\lambda(q), \rho_\lambda(q')) = \lambda\, d(q, q').$$

The balls that we consider here are defined via the Korányi metric

$$B_r^{\mathbb{H}}(q) = \{q' \in \mathbb{H};\ d(q, q') < r\} = q * B_r(0) = q * \rho_r(B_1(0)).$$

The vector fields

$$X = \partial_x - \frac{y}{2}\partial_z, \quad Y = \partial_y + \frac{x}{2}\partial_z, \quad Z = \partial_z$$

are left-invariant and form a basis for the Lie algebra of $\mathbb{H}$. The only non-trivial commutation relation is $Z = [X, Y]$

The horizontal gradient of a function $v : \mathbb{H} \to \mathbb{R}$ is given by

$$\nabla_{\mathbb{H}} v = (Xv)X + Y(v)Y,$$

while the horizontal second derivatives are given by the symmetric 2×2 matrix

$$D^{2,*}_{\mathbb{H}} v(q) = \begin{pmatrix} X^2 v(q) & \frac{XY+YX}{2} v(q) \\ \frac{XY+YX}{2} v(q) & Y^2 v(q) \end{pmatrix}.$$

The sub-elliptic or horizontal Laplacian is

$$\Delta_{\mathbb{H}} u = (X^2 + Y^2)u,$$

the sub-elliptic or horizontal p-Laplacian is

$$\Delta_{\mathbb{H},p} u = X(|\nabla_{\mathbb{H}} u|^{p-2} Xu) + Y(|\nabla_{\mathbb{H}} u|^{p-2} Yu),$$

and the normalized horizontal ∞-Laplacian is

$$\Delta_{\mathbb{H},\infty} u = \langle D^{2,*}_{\mathbb{H}} u \frac{\nabla_{\mathbb{H}} u}{|\nabla_{\mathbb{H}} u|}, \frac{\nabla_{\mathbb{H}} u}{|\nabla_{\mathbb{H}} u|} \rangle.$$

For a smooth function u, with non vanishing horizontal gradient, we define the normalized horizontal p-Laplacian as

$$\Delta^N_{\mathbb{H},p} u = |\nabla_{\mathbb{H}} u|^{2-p} \Delta_{\mathbb{H},p} u = \Delta_{\mathbb{H}} u + (p-2)\Delta_{\mathbb{H},\infty} u.$$

Let $\Omega \subset \mathbb{H}$ be a bounded domain. Fix $0 < \epsilon \leq 1$ and consider the ϵ-boundary strip given by

$$\Gamma^{\mathbb{H}}_{\epsilon} = \{x \in \mathbb{H} \setminus \Omega : \operatorname{dist}_{\mathbb{H}}(x, \Omega) \leq \epsilon\}.$$

Define $\Omega_\epsilon = \Omega \cup \Gamma^{\mathbb{H}}_{\epsilon}$ and denote by $\mathcal{B}(\Omega_\epsilon)$ and $\mathcal{B}(\overline{\Omega})$ the set of real-valued bounded measurable functions on Ω_ϵ and $\overline{\Omega}$, respectively. The natural p-mean average operator in the Heisenberg group

$$\mu^{\mathbb{H}}_p(\cdot, \epsilon) : \mathcal{B}(\Omega_\epsilon) \longrightarrow \mathcal{B}(\overline{\Omega})$$

is given by

$$\mu^{\mathbb{H}}_p(\phi, \epsilon)(y) = \mu_p^{\overline{B^{\mathbb{H}}_{\epsilon}(y)}}(\phi) \quad \text{for all } \phi \in \mathcal{B}(\Omega_\epsilon) \text{ and } y \in \overline{\Omega}.$$

The continuity, monotonicity, and affine invariance properties for $\mu_p^{\mathbb{H}}(\phi, \epsilon)$ follow as in the Euclidean case $\mu_p(\phi, \epsilon)$.

Rescaling and translating: Let $x \in \Omega$. Defining $u_{\epsilon,x}(z) = u(x * \rho_\epsilon(z))$ for $z \in B_1^{\mathbb{H}}(0)$, we have

$$\mu_p^{\mathbb{H}}(u, \epsilon)(x) = \mu_p^{\mathbb{H}}(u_{\epsilon,x}, 1)(0).$$

In the same way, we can associate a DPP to the natural p-means in $\mathbb{H}$

$$\begin{cases} u_\epsilon(x) = \mu_p^{\mathbb{H}}(u_\epsilon, \epsilon)(x) & \text{if } x \in \Omega, \\ u_\epsilon(x) = \quad G(x) & \text{if } x \in \Gamma_\epsilon^{\mathbb{H}}. \end{cases}$$

The natural p-means solutions $\mu_p^{\mathbb{H}}(u_\epsilon, \epsilon)$ to the $(DPP_\epsilon^{\mathbb{H}})$ satisfy the **UB** and the **CP**. The following lemma is an extension of the key lemma in [15]. This lemma has also been extended to the general case of Carnot groups in [1].

Lemma 1 ([1, 24], $1 < p < \infty$)

$$\textit{Let} \quad c_p = \frac{2}{(p+2)(p+4)} \left(\frac{\Gamma(\frac{p}{4} + \frac{3}{2})}{\Gamma(\frac{p}{4} + 1)} \right)^2,$$

where Γ is the Euler Gamma function. Let u be a smooth function in Ω_1 with $\nabla_{\mathbb{H}} u \neq 0$ in Ω_1. Then, we have the expansion

$$\mu_p^{\mathbb{H}}(u, \epsilon)(x_0) = u(x_0) + c_p \epsilon^2 \Delta_{\mathbb{H},p}^N u(x_0) + o(\epsilon^2) \quad \textit{as} \quad \epsilon \to 0$$

for all $x_0 \in \overline{\Omega}$.

Having established the correct asymptotic mean value formula, the next step in the convergence theorem. Unfortunately, the Euclidean method of Barles and Souganidis based on the strong uniqueness principle for the p-Laplacian $\Delta_{\mathbb{H},p}$ works only for Euclidean balls in $\mathbb{R}^3$, see [24]. To go from continuous to semi-discrete we build sub and super solutions of the $DPP_\epsilon^{\mathbb{H}}$ using the following perturbation lemma, valid for $p > 2$:

Lemma 2 ([12]) *Let Ω' be an open set containing Ω and let U be a function such that $\Delta_{\mathbb{H},p}^N U = 0$ and $\nabla_{\mathbb{H}} U \neq 0$ in Ω'. Then, there exist $\tilde{\epsilon} > 0$, $s \geq 4$ and $q_0 \in \mathbb{H}$ such that , denoting $v(x) = \|q_0^{-1} \star x\|_{\mathbb{H}}^s$, we have*

$$U + \epsilon v \textit{ is a sub-solution of } (DPP_\epsilon^{\mathbb{H}})$$

and

$$U - \epsilon v \textit{ is a supersolution of } (DPP_\epsilon^{\mathbb{H}})$$

in Ω_ϵ for all $0 < \epsilon < \tilde{\epsilon}$.

The key to prove this lemma is a strengthening of the expansion in ϵ of $\mu_p^{\mathbb{H}}$ that we are able to prove for $p > 2$.

Proposition 2 ([12]) *Let $p > 2$ and u be a smooth function in Ω_1 with $\nabla_{\mathbb{H}} u \neq 0$ in Ω_1. Then, there exist C and $\tilde{\epsilon} > 0$ such that*

$$|\mu_p^{\mathbb{H}}(u, \epsilon)(x_0) - u(x_0) - c_p \epsilon^2 \Delta_{\mathbb{H},p}^N u(x_0)| \leq C\epsilon^3$$

for all $0 < \epsilon < \tilde{\epsilon}$ and $x_0 \in \Omega$. In particular C and $\tilde{\epsilon}$ are uniform in $x_0 \in \Omega$ and depend only on p and the derivatives of u.

Note that we have replaced $o(\epsilon^2)$ by $O(\epsilon^3)$ when $p > 2$. As a consequence, the convergence when the boundary datum is the restriction of a p-harmonic function with a non-vanishing gradient follows.

Proposition 3 *Let Ω' be an open set containing Ω and let U be a function such that $\Delta_{\mathbb{H},p}^N U = 0$ and $\nabla_{\mathbb{H}} U \neq 0$ in Ω'. Denote by U_ϵ the solution of the $(DPP_\epsilon^{\mathbb{H}})$ with boundary datum U. Then*

$$U_\epsilon \to U \text{ uniformly in } \overline{\Omega}.$$

For arbitrary continuous boundary values, we need a regularity condition on the domain.

Definition 5 We say that a domain $\Omega \subset \mathbb{H}$ satisfies the exterior $\mathbb{H}$-corkscrew condition if there exists $\overline{\delta} > 0$ and $\mu \in (0, 1)$ such that for every $\delta \in (0, \overline{\delta})$ and $y \in \partial\Omega$ there exists a ball $B_{\mu\delta}^{\mathbb{H}}$ strictly contained in $B_\delta^{\mathbb{H}}(y) \setminus \Omega$.

There are $C^{1,\alpha}$ (in the Euclidean sense) domains that do not satisfy the exterior $\mathbb{H}$-corkscrew condition, while $C^{1,1}$ domains (in the Euclidean sense) which satisfy the exterior $\mathbb{H}$-corkscrew condition, see [8].

Theorem 8 (For the range $2 \leq p < \infty$) *Let Ω be an open, bounded subset of $\mathbb{H}$ satisfying the exterior $\mathbb{H}$-corkscrew condition with parameter $\mu \in (0, 1)$ and $G \in C(\Gamma_1)$. For $0 < \epsilon < 1$ let u_ϵ be the solution of the $(DPP_\epsilon^{\mathbb{H}})$ in Ω with boundary value G on Γ_ϵ. Given $\eta > 0$ there exist $\delta_0 = \delta_0(\eta, \mu, p)$ and $\epsilon_0 = \epsilon_0(\eta, \delta, \mu)$ such that*

$$|u_\epsilon(x) - G(y)| \leq \eta$$

for all $y \in \partial\Omega$, $x \in B_{\delta_0}^{\mathbb{H}}(y) \bigcap \Omega$ and all $\epsilon \leq \epsilon_0$.

The proof is done by iteration. It is an extension to $\mathbb{H}$ of the proof in [11] for the Euclidean case. Fix $\eta > 0$ and $y \in \partial\Omega$. The $\mathbb{H}$-corkscrew condition gives a sequence of balls $B_{\mu\delta_k}^{\mathbb{H}}(z_k)$ strictly contained in $B_{\delta_k}^{\mathbb{H}}(y) \setminus \Omega$ for all $k \in \mathbb{N}$. We prove the upper bound $u_\varepsilon(x) \leq G(y) + \eta$ for all $x \in B_{5\delta}^{\mathbb{H}}(y) \bigcap \Gamma_\epsilon^{\mathbb{H}}$ with the proof of the lower one being analogous (Fig. 1).

Proof (Sketch) We define the following sets

- $N = N^\epsilon(y) = N^{\epsilon,\delta}(y) := \sup_{B_{5\delta}^{\mathbb{H}}(y) \bigcap \Gamma_\epsilon} G$,
- $M = M^\epsilon(y) = \sup_{\Gamma_\epsilon} G$, and
- $M_k = M_k^\epsilon(y) = N^\epsilon(y) + \theta^k(M^\epsilon - N^\epsilon(y))$, for suitable $0 < \theta < 1$.

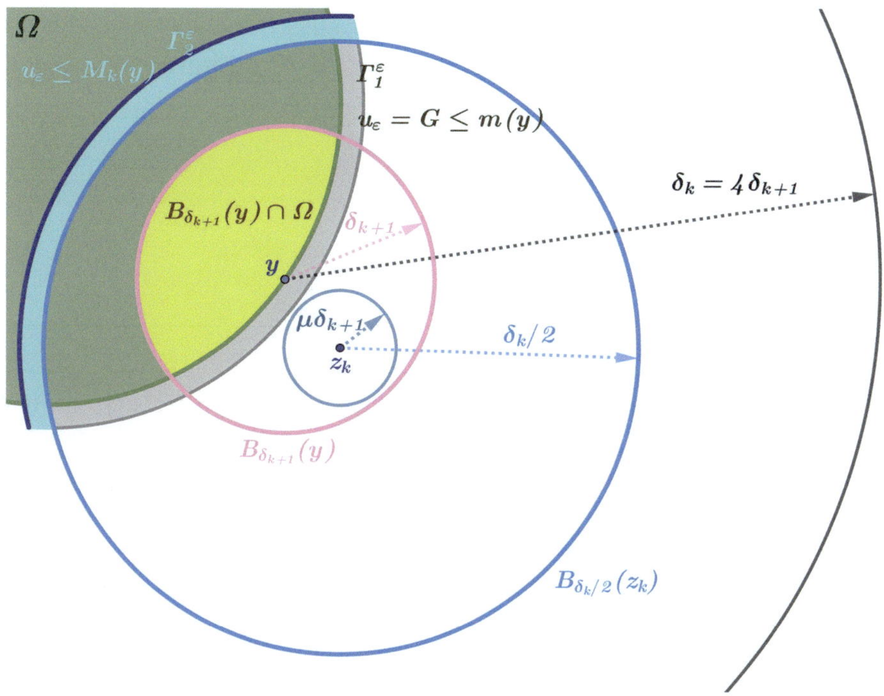

Fig. 1 This figure is taken from [11]

The following boundary value problem

$$\begin{cases} -\triangle^N_{\mathbb{H},p} U_k = 0 & \text{in } B^{\mathbb{H}}_{\delta_k}(z_k) \setminus \overline{B^{\mathbb{H}}_{\mu\delta_{k+1}}(z_k)}, \\ U_k = N & \text{on } \partial B^{\mathbb{H}}_{\mu\delta_{k+1}}(z_k), \\ U_k = M_k & \text{on } \partial B^{\mathbb{H}}_{\delta_k}(z_k) \end{cases}$$

has the radial solution

$$U_k(x) = \begin{cases} \dfrac{a_k}{|z_k^{-1} * x|_{\mathbb{H}}^{\frac{4-p}{p-1}}} & \text{if } p \neq 4, \\ a_k \log(|z_k^{-1} * x|_{\mathbb{H}}) + b_k & \text{if } p = 4. \end{cases}$$

Recall that the homogeneous dimension of $\mathbb{H}$ is 4.

The next steps are summarised as follows:

- In the annulus $B^{\mathbb{H}}_{\delta_k}(z_k) \setminus \overline{B^{\mathbb{H}}_{\mu\delta_{k+1}}(z_k)}$ we solve the $(DPP^{\mathbb{H}}_\epsilon)$ with boundary values U_k.
- We call the solution U_k^ϵ and observe that $\nabla U_k \neq 0$ in the annulus.
- We see that U_k^ϵ converges uniformly to U_k.

Using the comparison principle, the uniform convergence of Proposition 3 and a triangle inequality for the Koranyi norm, we have the following:

Claim: Let $\epsilon_k > 0$ and suppose that for all $\epsilon \in (0, \epsilon_k)$ we have

$$u_\epsilon \leq M_k^\epsilon(y) \text{ in } B_{\delta_k}(y) \bigcap \Omega.$$

Then, there exists $\epsilon_{k+1} \in (0, \epsilon_k)$ such that

$$u_\epsilon \leq M_{k+1}^\epsilon(y) \text{ in } B_{\delta_{k+1}}(y) \bigcap \Omega$$

for all $\epsilon \in (0, \epsilon_{k+1})$.

Using the claim above, since $u_\varepsilon \leq M^\varepsilon = M_0^\varepsilon$ in Ω_ε for all $\varepsilon \in (0, 1)$, we can find $\varepsilon_1 > 0$ such that $u_\varepsilon \leq M_1^\varepsilon(y)$ in $B_{\delta_1}^{\mathbb{H}}(y) \cap \Omega$ for $\varepsilon \leq \varepsilon_1$. We now repeat this process, and after k_0 iterations we find $\varepsilon_{k_0} > 0$ such that $u_\varepsilon \leq M_{k_0}^\varepsilon(y)$ in $B_{\delta_{k_0}}^{\mathbb{H}}(y) \cap \Omega$ for $\varepsilon \in (0, \varepsilon_{k_0})$. Choosing a suitable $k_0 \in \mathbb{N}$, we have

$$M_{k_0}^\varepsilon(y) - N^\varepsilon(y) = \theta^{k_0}(M^\varepsilon - N^\varepsilon(y)) \leq \frac{\eta}{2},$$

for all $y \in \partial\Omega$, because $M^\varepsilon \leq \sup_{\Gamma_\varepsilon^{\mathbb{H}}} G$ and $N^\varepsilon(y) \geq \inf_{\Gamma_\varepsilon^{\mathbb{H}}} G$. We finally get

$$u_\varepsilon \leq G(y) + \eta \quad \text{in} \quad B_{\delta_{k_0}}^{\mathbb{H}}(y) \cap \Omega,$$

for all $\varepsilon \in (0, \varepsilon_{k_0})$. □

We now study the convergence of u_ϵ as ϵ goes to zero, where u_ϵ is the solution of the $(DPP_\epsilon^{\mathbb{H}})$ with boundary value $G \in C(\Gamma_1)$. For $x \in \Omega$ define

$$\overline{u}(x) = \limsup_{\epsilon \to 0, q \to x} u_\epsilon(q) \quad \text{and} \quad \underline{u}(x) = \liminf_{\epsilon \to 0, q \to x} u_\epsilon(q).$$

Proposition 4 *The function $\overline{u}$ is a viscosity subsolution and the function $\underline{u}$ is a viscosity supersolution of $-\Delta_{\mathbb{H},p}^N u = 0$ in Ω.*

To conclude, use the convergence theorem

$$\limsup_{x \in \Omega \to x} \overline{u}(x) \leq G(y) \leq \liminf_{x \in \Omega \to x} \underline{u}(x)$$

and the comparison principle. Our main result is:

Theorem 9 (For $2 \leq p < \infty$) *Let Ω be an open bounded subset of $\mathbb{H}$ satisfying the exterior $\mathbb{H}$-corkscrew condition and let Γ_ϵ be its outer ϵ-boundary, for $0 < \epsilon < 1$. Let $g \in C(\partial\Omega)$ and $G \in C(\Gamma_1)$ be a continuous extension of g and let u_ϵ be the solution of $(DPP_\epsilon^{\mathbb{H}})$ with boundary datum G. Then u_ϵ converges to u uniformly on $\overline{\Omega}$ as $\epsilon \to 0$, where u is the solution to the Dirichlet problem*

$$\begin{cases} -\Delta_{\mathbb{H},p}^{N} u = 0 & in\ \Omega, \\ u = g & on\ \partial\Omega. \end{cases}$$

Acknowledgements I gratefully acknowledge the warming hospitality of the Ghent Analysis and PDE Center.

References

1. Adamovicz, T., Kijowski, A., Pinamonti, A., Warhurst , B.: Variational approach to the asymptotic mean value property for the p-Laplacian on Carnot Groups. Nonlinear Anal. **22**(198), 111893 (2020)
2. Armstrong, S., Smart, Ch.: An easy proof of jensen's theorem on the uniqueness of infinity harmonic functions. Calc. Var. PDE **37**, 381–384 (2010)
3. Aronsson, G.: Extension of functions satisfying lipschitz conditions. Arkiv Matematik **6**, 551–561 (1967)
4. Barles, G., Souganidis, P.: Convergence of approximation schemes for fully nonlinear second order equations. Asymptotic Anal. **4**(3), 271–283 (1991)
5. Bhattacharya, T., DiBenedetto, E., Manfredi, J.: Limits as $p \to \infty$ of $\Delta_p u = f$ and related extremal problems, Rendiconti del Seminario Matematico, pp. 15–68. Università e Politecnico Torino, Fascicolo Speciale Nonlinear PDE's (1989)
6. Blaschke, W.: Bin mittelwertsatz und eine kennzeichnende eigenschaft des logarithmischen potentials. Leipz. Ber. **68**, 3–7 (1916)
7. Bonfiglioli, A., Lanconelli, E.: Subharmonic functions in sub-Riemannian settings. J.E.M.S. **15**(2), 387–441 (2013)
8. Capogna, L., Garofalo, N.: Boundary behavior of nonnegative solutions of subelliptic equations in nta domains for carnot-carathéodory metrics. J. Fourier Anal. Appl. **4**(4–5), 403–432 (1998)
9. Chandra, E.W., Ishiwata, M., Magnanini, R., Wadade, H.: Variational p-harmonious functions: existence and convergence to p-harmonic functions. Nonlinear Differ. Equ. Appl. **28**(5), 51 (2021)
10. Crandall, M., Evans, L., Gariepy, R.: Optimal lipschitz extensions and the infinity laplacian. Calc. Var. PDE **13**, 123–129 (2001)
11. del Teso, F., Manfredi, J., Parviainen, M.: Convergence of dynamic programming principles for the p-laplacian. Adv. Calc. Var. **15**(2), 191–212 (2022)
12. Domokos, A., Manfredi, J., Ricciotti, D., Stroffolini, B.: Convergence of natural p-means for the p-Laplacian in the Heisenberg group. Nonlinear Anal. **223**, Paper No. 113058, 19 (2022)
13. Ferrari, F., Liu, Q., Manfredi, J.: On the characterization of p-harmonic functions on the heisenberg group by mean value properties. Discrete Continuous Dyn. Syst. Ser. A **34**(7), 2779–2793 (2014)
14. Heinonen, J., Kilpelainen, T., Martio, O.: Nonlinear potential theory of degenerate elliptic equations, pp. vi+363. Oxford Mathematical Monographs. Oxford Science Publications. The Clarendon Press, Oxford University Press, New York (1993)
15. Ishiwata, M., Magnanini, R., Wadade, H.: A natural approach to the asymptotic mean value property for the p-laplacian. Calc. Var. PDE **56**(4), 97 (2017)

16. Jensen, R.: Uniqueness of lipschitz extension: minimizing the sup norm of the gradient. Arch. Ration. Mech. Anal. **123**, 51–74 (1993)
17. Juutinen, P., Lindqvist, P., Manfredi, J.: On the equivalence of viscosity solutions and weak solutions for a quasi-linear equation. SIAM J. Math. Anal. **33**(3), 699–717 (2001)
18. Lewicka, M.: A course on tug-of-war games with random noise. In: Introduction and Basic Constructions, pp. ix+254. Universitext. Springer, Cham (2020)
19. Lewicka, M.: Noisy tug of war games for the p-laplacian: noisy tug of war games for the p-laplacian: Noisy tug of war games for the p-laplacian: Noisy tug of war games for the p-Laplacian: $1<p<\infty$. Indiana Univ. Math. J. **70**, 465–500 (2021)
20. Lewicka, M., Manfredi, J., Ricciotti, D.: Random walks and random tug-of-war in the heisenberg group. Math. Ann. **377**(1–2), 797–846 (2020)
21. Lindqvist, P.: On the definition and properties of p-superharmonic functions. J. Reine Angew. Math. **365**, 67–79 (1986)
22. Manfredi, J., Parviainen, M.,Rossi, J.: An asymptotic mean value characterization for p-harmonic functions. Proc. A.M.S. **138**(3), 881–889 (2010)
23. Manfredi, J., Parviainen, M., Rossi, J.: On the definition and properties of p-harmonious functions. Ann. Sc. Norm. Super. Pisa Cl. Sci. **11**(2), 215–241 (2012)
24. Manfredi, J., Stroffolini, B.: Convergence of the natural p-means for the p-Laplacian: ESAIM COCV. **27**(1), 1–18, (2021)
25. Peres, Y., Schramm, O., Sheffield, S., Wilson, D.B.: Tug-of-war and the infinity laplacian. J. Am. Math. Soc. **22**(1), 167–210 (2009)
26. Peres, Y., Sheffield, S.: Tug-of-war with noise: a game-theoretic view of the p-laplacian. Duke Math. J. **145**(1), 91–120 (2008)
27. Privaloff, I.: Sur les fonctions harmoniques. Rec. Math. Moscou (Mat. Sbornik) **32**, 464–471 (1925)
28. Saks, S.: On the Operators of Blaschke and Privaloff for subharmonic functions. Rec. Math. Moscou (Mat. Sbornik) **9**(2), 451–456 (1941)

Global Solutions for a Class of Nonlinear Evolution Equations in Supercritical Spaces $E^{\sigma,s}$

Baoxiang Wang and Xin Yu

Abstract We survey our recent works for a class of nonlinear evolution equations including Navier-Stokes, NLKG, nonlocal NLS and semilinear heat equation with initial data in super-critical spaces $E^{\sigma,s}$ for which the norm is defined as

$$\|u_0\|_{E^{\sigma,s}} = \|\langle\xi\rangle^{\sigma} 2^{s|\xi|}\widehat{u}_0\|_{L^2(\mathbb{R}^d)}, \ s < 0.$$

If $s < 0$, then $H^{\kappa} \subset E^{\sigma,s}$ for any $\kappa, \sigma \in \mathbb{R}$. The global existence and uniqueness of solutions are obtained for the initial data $u_0 \in E^{\sigma,s}$ and $\widehat{u}_0$ is supported in the first octant, where $s < 0$, σ is the (sub-)critical index for the nonlinear evolution equations.

Keywords Function space · Navier–Stokes equation · NLKG · Nonlocal NLS · Semilinear heat equation

MSC2020 35K55 · 35L70 · 35Q30 · 35Q55 · 42B35

1 Introduction

Generally speaking, for a nonlinear evolution equation (NLE), there exists a *critical Sobolev space* H^{σ_c} such that it is globally well-posed in H^{σ} for $\sigma > \sigma_c$ and ill-posed in H^{σ} for $\sigma < \sigma_c$. H^{σ} with $\sigma < \sigma_c$ is said to be a *super-critical Sobolev space* for the NLE. So, for the general initial data, it is impossible to get the global well-posedness of solutions for NLE in super-critical Sobolev spaces. A Banach function space X defined on $\mathbb{R}^d$ is said to be a *super-critical space* to the NLE, if there exists $\sigma < \sigma_c$ such that $H^{\sigma} \subset X$.

For example, the Navier-Stokes equation (NS) in $\mathbb{R}^d$

$$u_t - \Delta u + u \cdot \nabla u + \nabla p = 0, \ \ \operatorname{div} u = 0, \ \ u(0, x) = u_0$$

B. Wang (✉) · X. Yu
School of Sciences, Jimei University, Xiamen, P. R. of China
e-mail: wbx@jmu.edu.cn

M. Chatzakou et al. (eds.), *Direct and Inverse Problems with Applications*,
Trends in Mathematics 12, https://doi.org/10.1007/978-3-031-98645-1_14

has the critical Sobolev space $H^{d/2-1}$, which is locally well-posed in H^σ ($\sigma \geq d/2-1$), and globally well-posed in H^σ ($\sigma \geq d/2-1$) with sufficiently small norm in $H^{d/2-1}$. However, it is ill-posed in H^σ for any $\sigma < d/2-1$. The solution of NS satisfies the following conservation law:

$$\frac{1}{2}\|u(t)\|^2_{L^2(\mathbb{R}^d)} + \int_0^t \|\nabla u(s)\|^2_{L^2(\mathbb{R}^d)} ds = \frac{1}{2}\|u_0\|^2_{L^2(\mathbb{R}^d)}.$$

Up to now, we do not know if there are some other conservation laws or "a priori" estimates in H^σ (say for some $\sigma \geq d/2-1$) of NS, so the global well-posedness in H^σ ($\sigma \geq d/2-1$) with large data is open. Noticing that L^2 is a super-critical space for NS, it is impossible to obtain the global well-posedness result in L^2. In fact, one can only get the existence of global weak solutions of NS in L^2, but the uniqueness of the weak solution in L^2 is still open.

We may further ask if there are some other super-critical spaces X so that NS has the global well-posedness in X. By introducing a class of super-critical function spaces $E^{\sigma,s}$ with $s < 0$, which is rougher than any Sobolev spaces H^κ, we will show the global existence and uniqueness of solutions for Navier-Stokes, NLKG, nonlocal NLS and semi-linear heat equations, our main assumption on initial data is that their Fourier transforms are supported in the first octant or a cone-like domain.

1.1 Function Spaces $E^{\sigma,s}$

Let $\mathcal{S}$ be the Schwartz space and $\mathcal{S}'$ be its dual space. We write $|x| = |x_1| + \cdots + |x_d|$ for $x = (x_1, ..., x_d) \in \mathbb{R}^d$ and

$$p_\lambda(f) = \sup_{x\in\mathbb{R}^d} e^{\lambda|x|}|f(x)|, \quad q_\lambda(f) = \sup_{\xi\in\mathbb{R}^d} e^{\lambda|\xi|}|\widehat{f}(\xi)|,$$

$$\mathcal{S}_1 := \{f \in \mathcal{S} : p_\lambda(f) + q_\lambda(f) < \infty, \ \forall\, \lambda > 0\}.$$

$\mathcal{S}_1$ equipped with the system of semi-norms $\{p_\lambda + q_\lambda\}_{\lambda>0}$ is a complete locally convex linear topological space, which is said to be the Gelfand-Shilov space, cf. [5]. We denote by $\mathcal{S}_1'$ the dual space of $\mathcal{S}_1$. One easily sees that

$$\mathcal{S}_1 \subset \mathcal{S}, \ \ \mathcal{S}' \subset \mathcal{S}_1'.$$

$\mathcal{S}_1$ contains the translations and modulations of Gaussian $e^{-imx}e^{-|x-n|_2^2/2}$ and their linear combinations, where $|x|_2 = (x_1^2 + \cdots + x_d^2)^{1/2}$ for $x = (x_1, ..., x_d)$. The Fourier transforms on $\mathcal{S}_1'$ can be defined by duality (cf. [6]). For $s, \sigma \in \mathbb{R}$, we denote

$$E^{\sigma,s} := \{f \in \mathcal{S}_1' : \langle\xi\rangle^\sigma 2^{s|\xi|}\widehat{f}(\xi) \in L^2(\mathbb{R}^d)\}$$

for which the norm is defined by $\|f\|_{E^{\sigma,s}} = \|\langle\xi\rangle^{\sigma} 2^{s|\xi|} \widehat{f}(\xi)\|_{L^2}$. $E^{\sigma,s}$ is a generalization of Sobolev spaces $H^{\sigma} := (I - \Delta)^{-\sigma/2} L^2(\mathbb{R}^d)$ for which the norm is defined by

$$\|f\|_{H^{\sigma}} := \|\langle\xi\rangle^{\sigma} \widehat{f}\|_{L^2}.$$

One easily sees that $H^{\sigma} = E^{\sigma,0}$. In the case $s > 0$, $E^s := E^{0,s}$ was introduced by Björck [1] and it is a smooth function space. The spaces like $E^{\sigma,s}$ ($s > 0$) have important applications in studying the Gevrey regularity for the solutions of evolution equations (see e.g. Foias and Temam [7], Bourdaud et al. [2]). However, in the case $s < 0$, $E^{\sigma,s}$ is much rougher than any H^{κ}, and there is no more study in the literature. We have the following inclusion (cf. [4]),

$$H^{\kappa} \subset E^{\sigma,s}, \quad \forall\, \kappa, \sigma \in \mathbb{R},\ s < 0. \tag{1}$$

By (1), we see that $\cup_{\kappa\in\mathbb{R}} H^{\kappa}$ is a subset of $E^{\sigma,s}$ if $s < 0$.

1.2 Examples

In 1D case, let $\delta(x)$ be the Dirac measure and

$$F(x) = \left(\mathrm{i}\delta(x) + p.v.\frac{1}{x}\right)$$

be the Sokhotski-Plemelj distribution, $\mathrm{i}\widehat{F}/2$ is the Heaviside function. For any $k \in \mathbb{N}$, we have $\partial^k F \in E^{\sigma,s}$ and supp $\widehat{F} \subset [0, \infty)$ if $s < 0$. In higher spatial dimensions, we see that the distribution

$$G(x) := \partial^{\alpha} \prod_{j=1}^{d} F(x_j) \in E^{\sigma,s}, \quad \operatorname{supp} \widehat{G} \subset \mathbb{R}^d_I,$$

where

$$\mathbb{R}^d_I := \{\xi \in \mathbb{R}^d : \xi_j \geq 0,\ j = 1, ..., d\}.$$

2 Semi-linear Heat Equation in $E^{\sigma,s}$

In view of (1), $E^{\sigma,s}$ with $s < 0$ is a super-critical space for any nonlinear evolution equation if it has a critical Sobolev space H^{σ_c}. Let us consider the semi-linear heat equation (SLH)

$$\partial_t u - \Delta u - u^m = 0, \quad u(0, x) = u_0(x) \tag{2}$$

in $E^{\sigma,s}$ with $s < 0$, where u is a complex-valued distribution function of $(t, x) \in \mathbb{R}_+ \times \mathbb{R}^d$, $m \geq 2$ is a natural number.

For any $k \in \mathbb{Z}^d$, we denote

$$\Box_k = \mathcal{F}^{-1} \chi_{k+[0,1)^d} \mathcal{F},$$

where χ_A denote the characteristic function on $A \subset \mathbb{R}^d$. For convenience, we denote for any $\mathbb{A} \subset \mathbb{Z}^d$,

$$\|u\|_{\widetilde{L}^\gamma(0,\infty;E^{\sigma,s}_{p,q}(\mathbb{A}))} = \left(\sum_{k\in\mathbb{A}} 2^{s|k|q} \langle k \rangle^{\sigma q} \|\Box_k u\|^q_{L^\gamma_t L^p_x(\mathbb{R}_+\times\mathbb{R}^d)} \right)^{1/q}$$

and $\widetilde{L}^\gamma(0,\infty; E^{\sigma,s}_{p,q}) := \widetilde{L}^\gamma(0,\infty; E^{\sigma,s}_{p,q}(\mathbb{Z}^d))$, $\widetilde{L}^\gamma(0,\infty; E^{s}_{p,q}) := \widetilde{L}^\gamma(0,\infty; E^{0,s}_{p,q})$. We have the following (cf. [4])

Theorem 1 *Let $d \geq 1$, $m \in \mathbb{N} \setminus \{1\}$. Let $s \leq 0$, $\sigma \geq d/2 - 2/(m-1)$, $u_0 \in E^{\sigma,s}$ with* supp $\widehat{u}_0 \subset \mathbb{R}^d_I \setminus \{0\}$. *We have the following results.*

- *If $s < 0$, then there exists $s_0 \leq s$ such that SLH (2) has a unique solution $u \in \widetilde{L}^m(0,\infty; E^{\sigma+2/m,s_0}_{2,2})$ satisfying the integral equation*

$$u(t) = e^{t\Delta}u_0 + \int_0^t e^{(t-\tau)\Delta}u^m(\tau)\, d\tau \tag{3}$$

  and

$$u \in C([0,\infty); E^{\sigma,s_0}) \cap \widetilde{L}^\infty(0,\infty; E^{\sigma,s_0}_{2,2}) \cap \widetilde{L}^1(0,\infty; E^{2+\sigma,s_0}_{2,2}). \tag{4}$$

  Moreover, condition supp $\widehat{u}_0 \subset \mathbb{R}^d_I$ *is sharp in the sense that, for any $s_0 \leq s$, the solution map $u_0 \to u$ from $E^{\sigma,s}$ into E^{σ,s_0} is not C^m for certain initial data whose Fourier transforms are supported in $\mathbb{R}^d_I \cup (-\mathbb{R}^d_I)$.*
- *Let $s = 0$. Assume that $\|u_0\|_{E^{\sigma,0}}$ is sufficiently small, then (3) has a unique solution u satisfying (4) for $s_0 = 0$. Moreover, $\sigma \geq \sigma_c = d/2 - 2/(m-1)$ is optimal in the sense that the solution map $u_0 \to u$ is not C^m in $E^{\sigma,0}_+ := \{f \in E^{\sigma,0} :$* supp $\widehat{f} \subset \mathbb{R}^d_I \setminus \{0\}\}$ *if $\sigma < \sigma_c$.*

It is worth to mention that in the 1D case, $m = 2$ is less than the Fujita's critical power $1 + 2/d = 3$, and SLH has the blowup solutions for small data. In Theorem 1, we obtain a new global well-posedness result in critical Sobolev space for $m = 2$ in 1D.

3 NLKG in $E^{\sigma,s}$

We consider the Cauchy problem for the complex nonlinear Klein-Gordon (NLKG) and sinh-Gordon equations

$$\begin{cases} \partial_t^2 u + u - \Delta u + f(u) = 0, \\ u(x,0) = u_0(x), \; u_t(x,0) = u_1(x), \end{cases} \tag{5}$$

where $u(x,t)$ is a complex-valued function of $(x,t) \in \mathbb{R}^{d+1}$ and $f(u) = \pm u^{1+\alpha}$, $\alpha \in \mathbb{N}$ or $f(u) = \sinh u - u$, $\sinh u = (e^u - e^{-u})/2$.

Let ψ be a smooth radial bump function satisfying $\psi(\xi) = 1$ for $|\xi| \leqslant 1$ and $\operatorname{supp} \psi = \{\xi : |\xi| \leqslant 5/4\}$. Denote $\varphi(\xi) = \psi(\xi) - \psi(2\xi)$ and $\varphi_j(\xi) = \varphi(2^{-j}\xi)$. One sees that

$$\psi(\xi) + \sum_{j \geqslant 1} \varphi_j(\xi) = 1, \quad \forall\, \xi \in \mathbb{R}^d$$

and

$$\sum_{j \in \mathbb{Z}} \varphi_j(\xi) = 1, \quad \forall\, \xi \in \mathbb{R}^d \setminus \{0\}.$$

We write

$$\triangle_0 = \mathcal{F}^{-1} \psi \mathcal{F}, \quad \triangle_j = \mathcal{F}^{-1} \varphi_j \mathcal{F}, \quad j \in \mathbb{N}$$

For our purpose, we introduce the following resolution spaces

$$\|u\|_{\widetilde{L}^{\gamma}(\mathbb{R}; B^{\sigma,s}_{p,q})} = \left(\sum_{j=0}^{\infty} 2^{\sigma j q} \|2^{s|\nabla|} \triangle_j u\|^q_{L^{\gamma}_t L^p_x(\mathbb{R} \times \mathbb{R}^d)} \right)^{1/q}.$$

Denote $\|f\|_{H^{\sigma,s}_p} = \|\mathcal{F}^{-1} \langle \xi \rangle^{\sigma} 2^{s|\xi|} \mathcal{F} f\|_{L^p}$. We have the following results (cf. [8]):

Theorem 2 *Let $d \geqslant 1$, $f(u) = u^{1+\alpha}$, $4/d \leqslant \alpha < \infty$, $\alpha \in \mathbb{N}$, $s < 0$ and $\sigma \geqslant \max(\sigma_c, 1/2)$. Suppose that $(u_0, u_1) \in E^{\sigma,s} \times E^{\sigma-1,s}$ with* $\operatorname{supp} \widehat{u}_0$, $\operatorname{supp} \widehat{u}_1 \subset \mathbb{R}^d_I \setminus \{0\}$ *for $d \leqslant 6$, and* $\operatorname{supp} \widehat{u}_0$, $\operatorname{supp} \widehat{u}_1 \subset \mathbb{R}^d_I$ *for $d \geqslant 7$. Then there exists $s_0 \leqslant s$ such that NLKG (5) has a unique mild solution $u \in C(\mathbb{R}; E^{\sigma,s_0}) \cap X^{\sigma,s_0}_p$, where the norm on $X^{\sigma,s}_p$ is defined by*

$$\|u\|_{X^{\sigma,s}_p} = \begin{cases} \|u\|_{\widetilde{L}^{\infty}(\mathbb{R}, B^{\sigma,s}_{2,2})} + \|u\|_{\widetilde{L}^{p}(\mathbb{R}, B^{\sigma-1/2,s}_{p,2})}, & \sigma > 1/2; \\ \|u\|_{L^{\infty}(\mathbb{R}, E^{1/2,s})} + \|u\|_{L^{p}(\mathbb{R}, H^{0,s}_p)}, & \sigma = 1/2, \end{cases}$$

where

$$p = \begin{cases} 2(d+1)/(d-1), & \text{for } \alpha \geqslant 4/(d-1); \\ 2(d+1+\theta)/(d-1+\theta), & \text{for } \alpha = 4/(d-1+\theta),\ \theta \in (0,1]. \end{cases}$$

Since $H^1 \times L^2 \subset E^{\sigma,s} \times E^{\sigma-1,s}$, $s < 0$, we see that for $4/(d-2) < \alpha < \infty$, $(u_0, u_1) \in H^1 \times L^2$, we obtain the uniqueness of the mild solution of NLKG in the condition $\operatorname{supp} \widehat{u}_0$, $\operatorname{supp} \widehat{u}_1 \subset \mathbb{R}^d_I \setminus \{0\}$.

Theorem 3 *Let* $d \geqslant 2$, $f(u) = \sinh u - u$, $s < 0, \sigma \geqslant d/2$, $(u_0, u_1) \in E^{\sigma,s} \times E^{\sigma-1,s}$ *with* $\operatorname{supp} \widehat{u}_0$, $\operatorname{supp} \widehat{u}_1 \subset \mathbb{R}^d_I \setminus \{0\}$ *for* $d \leqslant 4$, *and* $\operatorname{supp} \widehat{u}_0$, $\operatorname{supp} \widehat{u}_1 \subset \mathbb{R}^d_I$ *for* $d \geqslant 5$. *Then there exists* $s_0 \leqslant s$ *such that NLKG (5) has a unique mild solution* $u \in C(\mathbb{R}; E^{\sigma,s_0}) \cap Y^{\sigma,s_0}_r$, *where the norm on* $Y^{\sigma,s}_r$ *is defined by*

$$\|u\|_{Y^{\sigma,s}_r} = \|u\|_{\widetilde{L}^\infty(\mathbb{R}, B^{\sigma,s}_{2,2})} + \|u\|_{\widetilde{L}^r(\mathbb{R}, B^{\sigma-1/2,s}_{r,2})},$$

where $r = 2 + 4/(d-1)$ *for* $d \geqslant 3$; $r = 2 + 4/d$ *for* $d = 2$.

4 Navier–Stokes Equation in $E^s_{r,1}$

We consider the Cauchy problem for the incompressible Navier-Stokes equation (NS):

$$u_t - \Delta u + u \cdot \nabla u + \nabla p = 0, \quad \operatorname{div} u = 0, \quad u(0, x) = u_0, \tag{6}$$

where $u = (u_1, ..., u_d)$ denotes the flow velocity vector, and p describes the scalar pressure, $u_0 = (u^1_0, ..., u^d_0)$ is an initial velocity with $\operatorname{div} u_0 = 0$. $u_t = \partial u/\partial t$, Δ denotes the Laplacian. It is easy to see that (6) can be rewritten as the following equivalent form:

$$u_t - \Delta u + \mathbb{P}\operatorname{div}(u \otimes u) = 0, \quad u(0, x) = u_0,$$

where $\mathbb{P} = I - \operatorname{div} \Delta^{-1}\nabla$ is the matrix operator projecting onto the divergence-free vector fields, I is the identity matrix.

For $s \in \mathbb{R}$, $1 < p, q < \infty$, we write

$$\|f\|_{E^s_{p,q}} = \|\{2^{s|k|}\Box_k f\}\|_{\ell^q(L^p)}.$$

We have the following result (see [6]):

Theorem 4 *Let* $d \geq 2$, $s < 0$, $2 \leq r \leq d$. *Assume that* $u_0 \in E^s_{r,1}$ *and* $\operatorname{supp} \widehat{u}_0 \subset \mathbb{R}^d_I$. *Then there exists* $s_0 \leq s$ *such that NS has a unique mild solution* $u \in \widetilde{L}^\gamma(0, \infty; E^{s_0}_{p,1}) \cap C(0, \infty; E^{s_0}_{r,1})$, *where*

$$(\gamma, p) = \begin{cases} (d+2, d+2), & r = d \\ (2, \rho), & 2 \leq r < d \end{cases}$$

for some sufficiently large $\rho := \rho(r)$ *and*

$$\|u\|_{\widetilde{L}^{\gamma}(I,E^{s}_{p,1})}=\left\|\{2^{s|k|}\|\Box_k u\|_{L^{\gamma}_{t\in I}L^{p}_{x}}\}_{k\in\mathbb{Z}^d}\right\|_{\ell^1_k}.$$

Moreover, there exists a small constant $c>0$ such that the solution is real analytic when $t>s^2/c^2$.

Since $L^2\subset E^{s}_{r,1}$ for $2\le r\le d$, we see that for $u_0\in L^2$ and $\operatorname{supp}\widehat{u}_0\subset\mathbb{R}^d_I$, we have shown the uniqueness of the mild solution of NS.

5 Nonlocal NLS in $E^{\sigma,s}$

The nonlocal nonlinear Schrödinger equation (NNLS) takes the form

$$\mathrm{i}u_t+\partial_x^2u+\alpha\, u^2u^*=0,\ \ u(0,x)=u_0(x), \tag{7}$$

where $u(t,x)$ is a complex valued function of $(t,x)\in[0,T)\times\mathbb{R}$ for some $0<T\le\infty$, $u^*(t,x)=\bar{u}(t,-x)$ and $\bar{u}$ stands for the conjugate complex number of u, $\alpha\in\mathbb{R}$.

Theorem 5 *Let $s\le 0$, $-1/2<\sigma\le 0$, $u_0\in E^{\sigma,s}$ with $\operatorname{supp}\widehat{u}_0\subset[0,\infty)$. Then there exists a $T>0$ such that NNLS (7) has a unique mild solution $u\in L^\infty(0,T;E^{\sigma,s})\cap X^{s}_{\sigma,\Delta,T}$ satisfying*

$$u(t)=e^{\mathrm{i}t\Delta}u_0-\mathrm{i}\int_0^t e^{\mathrm{i}(t-\tau)\Delta}(u^2u^*)(\tau)d\tau,$$

where $X^{s}_{\sigma,\Delta,T}$ is a resolution space. Moreover, if $s<0$, $\operatorname{supp}\widehat{u}_0\subset[\varepsilon_0,\infty)$ for some $\varepsilon_0>0$, then there exists some $j_0:=j_0(\alpha,s,\varepsilon_0,\sigma,\|u_0\|_{E^{\sigma,s}})\in\mathbb{N}$ such that NNLS has a unique mild global solution

$$u\in L^\infty(0,2^{\sqrt{j}};E^{sj}_{\sigma})\cap X^{sj}_{\sigma,\Delta,2^{\sqrt{j}}},\ \ \forall\, j\ge j_0$$

satisfying

$$\sup_{j\ge j_0}\|u\|_{L^\infty(0,2^{\sqrt{j}};E^{sj}_{\sigma})\cap X^{sj}_{\sigma,\Delta,2^{\sqrt{j}}}}<\infty.$$

The above method is also adapted to the derivative version of NNLS:

$$\mathrm{i}u_t+\partial_x^2u+\alpha\, uu^*\partial_xu+\beta\, u^2\partial_xu^*=0,\ \ u(0,x)=u_0(x)$$

and one can find the details of the proof of Theorem 5 in [3].

Acknowledgements The authors were supported in part by NSFC Grant Number 12171007 and 12450001. The first named author is very grateful to Professor M. Ruzhansky for his invitation to visit Ghent University.

References

1. Björck, G.: Linear partial differential operators and generalized distributions. Ark. Mat. **6**, 351–407 (1966)
2. Bourdaud, G., Reissig, M., Sickel, W.: Hyperbolic equations, function spaces with exponential weights and nemytskij operators. Ann. Mat. Pura Appl. **182**, 409–455 (2003)
3. Chen, J., Lu, Y.F., Wang, B.X.: Global cauchy problems for the nonlocal (derivative) nls in e^s_σ. J. Differ. Equ. **344**, 767–806 (2023)
4. Chen, J., Wang, B.X., Wang, Z.M.: Complex valued semi-linear heat equations in super-critical spaces $e^{\sigma,s}$. Math. Ann. **386**, 1351–1389 (2023)
5. Gelfand, I.M., Shilov, G.E.: Generalized Functions, Spaces of Fundamental and Generalized Functions, vol. 2. Academic, New York-London (1968)
6. Feichtinger, H., Gröchenig, K., Li, K.J., Wang, B.X.: Navier-stokes equation in super-critical spaces $e^s_{p,q}$. Ann. Inst. H. Poincaré Anal. Non Linéaire **38**(1), 139–173 (2021)
7. Foias, C., Temam, R.: Gevrey class regularity for the solutions of the Navier–Stokes equations. J. Funct. Anal. **87**(2), 359–369 (1989)
8. Wang, B.X.: Global Cauchy problem for the complex NLKG and sinh-Gordon equations in super-critical spaces. J. Funct. Anal. **287**(2), 110458, 1–64 (2024)

An Index Transform Method for Solutions of the Boundary Value Problems in a Wedge

Semyon Yakubovich

Abstract We exhibit applications of the index transforms recently established by the author, which involve modified Bessel and associated Legendre functions as the kernels to solve boundary value problems in a wedge for higher order PDE's. This class of transformations has a feature in integration with respect to a parameter (index) of the special function. The classical and basic examples are the Kontorovich-Lebedev and Mehler-Fock transforms. Concrete examples are demonstrated.

Keywords Index transforms · Kontorovich-Lebedev transform · Modified bessel function · Kelvin functions · Associated legendre functions

MSC2020: 44A15 · 41A60 · 33C10 · 33C70 · 35A25 · 35G15

1 Preliminaries

The reciprocal pair of the Kontorovich-Lebedev transforms [11]

$$g(\tau) = \int_0^\infty K_{i\tau}(\lambda x) f(x) dx, \quad \tau > 0,$$

$$f(x) = \frac{2}{\pi^2 x} \int_0^\infty \tau \sinh(\pi\tau) K_{i\tau}(\lambda x) g(\tau) d\tau, \quad \lambda > 0, \tag{1}$$

has been used in many applications, including fluid mechanics and quantum optics. For instance (cf. [12]), considering the Laplace equation in cylindrical coordinates

$$\frac{\partial^2 u}{\partial r^2} + \frac{1}{r}\frac{\partial u}{\partial r} + \frac{1}{r^2}\frac{\partial^2 u}{\partial \theta^2} + \frac{\partial^2 u}{\partial z^2} = 0, \tag{2}$$

S. Yakubovich (✉)
Department of Mathematics, Faculty of Sciences, University of Porto, Porto, Portugal
e-mail: syakubov@fc.up.pt

M. Chatzakou et al. (eds.), *Direct and Inverse Problems with Applications*,
Trends in Mathematics 12, https://doi.org/10.1007/978-3-031-98645-1_15

we seek a solution in the form $u(r, \theta, z) = e^{\pm i\theta \pm i\sigma z} R(r)$ which yields for $R(r)$ Bessel's differential equation

$$R'' + \frac{1}{r} R' + \left(\frac{\tau^2}{r^2} - \sigma^2 \right) R = 0.$$

One of its solutions is the modified Bessel function $K_{i\tau}(|\sigma| r)$. Therefore, $u(r, \theta, z) = e^{\pm i\theta \pm i\sigma z} K_{i\tau}(|\sigma| r)$. Hence, we are looking for a solution of the form $u(r, \theta, z) = e^{i\sigma z} u(r, \theta)$ in the wedge

$$W \equiv \{(r, \theta) : \ 0 < r < \infty, \ 0 \leq \theta \leq \alpha\}$$

subject to the conditions

$$u(r, 0) = 0, \qquad u(r, \alpha) = r f(r) \tag{3}$$

under a suitable class of functions f. Then, in a straightforward way of differentiation under the integral sign, one verifies the solution of the boundary value problem (3) for the equation

$$\frac{\partial^2 u}{\partial r^2} + \frac{1}{r} \frac{\partial u}{\partial r} + \frac{1}{r^2} \frac{\partial^2 u}{\partial \theta^2} - \lambda^2 u = 0,$$

in terms of the inverse Kontorovich-Lebedev transform (1)

$$u(r, \theta) = \frac{2}{\pi^2} \int_0^\infty \frac{\tau \sinh(\pi \tau)}{\sinh(\alpha \tau)} K_{i\tau}(\lambda r) g(\tau) \sinh(\theta \tau) d\tau. \tag{4}$$

The general solution of the Laplace equation (2) can be written in terms of the iterated integral due to the superposition principle

$$u(r, \theta, z) = \frac{2}{\pi^2} \int_{-\infty}^{\infty} e^{i\sigma z} d\sigma \int_0^\infty \frac{\tau \sinh(\pi \tau)}{\sinh(\alpha \tau)} K_{i\tau}(|\sigma| r) g(\tau) \sinh(\theta \tau) d\tau.$$

Interchanging the order of integration under suitable conditions, we calculate the known Fourier integral with respect to σ for the modified Bessel function, rewriting solution (4) in the form

$$u(r, \theta, z) = \frac{4}{\pi \sqrt{z^2 + r^2}} \int_0^\infty \frac{\tau \sinh(\pi \tau / 2) \sinh(\theta \tau)}{\sinh(\alpha \tau)} \times \cos\left(\tau \log\left(\frac{z}{r} + \sqrt{\frac{z^2}{r^2} + 1} \right) \right) g(\tau)\, d\tau.$$

2 Higher Order PDEs

In this section, we extend the method of the index transforms to solve boundary value problems in the wedge (3) for other PDEs of the order ≥ 2. Indeed, let us consider the reciprocal index transforms (cf. [4])

$$g(\tau) = \frac{\cosh(\pi\tau)}{\sqrt{\pi}} \int_0^\infty \left[\frac{\tau e^{x/2}}{x \sinh(\pi\tau)} \mathrm{Re}\left[I_{i\tau}\left(\frac{x}{2}\right)\right] - \frac{1}{2\pi(1+\tau^2)} {}_2F_2\left(1, \frac{3}{2}; 2-i\tau, 2+i\tau; x\right)\right] f(x)\, dx, \quad \tau \in \mathbb{R},$$

$$f(x) = \sqrt{\pi} e^{-x/2} \int_0^\infty \mathrm{Re}\left[I_{i\tau}\left(\frac{x}{2}\right)\right] \frac{g(\tau) d\tau}{\cosh(\pi\tau)}, \quad x > 0,$$

where ${}_2F_2(a, b; c, d; z)$ is the generalized hypergeometric function, $I_\nu(z)$ is the modified Bessel function of the first kind [1], and the real part is taken with respect to the pure imaginary index $i\tau$. It is proved in [4] that the kernel

$$\frac{\sqrt{\pi}}{\cosh(\pi\tau)} e^{-x/2} \mathrm{Re}\left[I_{i\tau}\left(\frac{x}{2}\right)\right]$$

satisfies the following second-order differential equation by x with polynomial coefficients

$$x^2 u'' + x(1+x)u' + \left(\frac{x}{2} + \tau^2\right) u = 0.$$

Hence, one can construct a solution to the boundary value problem in the wedge (3) for the second-order partial differential equation

$$\sqrt{x^2+y^2} \left(\frac{\partial^2 u}{\partial x^2} + \frac{\partial^2 u}{\partial y^2} \right) + x \frac{\partial u}{\partial x} + y \frac{\partial u}{\partial y} + \frac{1}{2} u = 0, \ (x, y) \in \mathbb{R}^2 \backslash \{0\}. \tag{5}$$

In fact, writing (2) in polar coordinates, we end up with the equation

$$r^2 \frac{\partial^2 u}{\partial r^2} + r(1+r) \frac{\partial u}{\partial r} + \frac{\partial^2 u}{\partial \theta^2} + \frac{r}{2} u = 0.$$

Its solution is the integral

$$u(r, \theta) = \sqrt{\pi}\, e^{-r/2} \int_0^\infty \mathrm{Re}\left[I_{i\tau}\left(\frac{r}{2}\right)\right] \frac{g(\tau) \sinh(\theta\tau)}{\sinh(\alpha\tau) \cosh(\pi\tau)} d\tau, \quad r > 0,$$

under the boundary conditions

$$u(r, 0) = 0, \qquad u(r, \alpha) = f(r). \tag{6}$$

Further, it is proved in [7] that the kernel, involving the square of the associated Legendre function of the first kind [1]

$$\sqrt{\pi}\Gamma\left(\frac{1}{2}+i\tau-\mu\right)\Gamma\left(\frac{1}{2}-i\tau-\mu\right)\left[P^{\mu}_{-1/2+i\tau}\left(\sqrt{\frac{x+1}{x}}\right)\right]^2,$$

where $\Gamma(z)$ is the Euler gamma function, is a solution of the third order differential equation with polynomial coefficients

$$2x^3(1+x)\frac{d^3u}{dx^3}+3x(1+2x)\frac{d^2u}{dx^2}+x\left(2x(1-\mu^2)+2\tau^2+\frac{1}{2}\right)\frac{du}{dx}-\left(\frac{1}{4}+\tau^2\right)u=0.$$

Consequently, the third-order PDE

$$\begin{aligned}&x(r(1+r)-y^2)\frac{\partial^3u}{\partial x^3}+y(r(1+r)-x^2)\frac{\partial^3u}{\partial y^3}\\&+x(r(1+r)-x^2+2y^2)\frac{\partial^3u}{\partial x\partial y^2}+y(r(1+r)-y^2+2x^2)\frac{\partial^3u}{\partial y\partial x^2}\\&+\left(\frac{3}{2r}+2\right)\left(\left(y^2+\frac{x^2}{r}\right)\frac{\partial^2u}{\partial x^2}+\left(x^2+\frac{y^2}{r}\right)\frac{\partial^2u}{\partial y^2}\right)\\&+\frac{(3+r)xy}{r^2}\frac{\partial^2u}{\partial x\partial y}-\left(\mu^2-1+\frac{1}{4r}\right)\left(x\frac{\partial u}{\partial x}+y\frac{\partial u}{\partial y}\right)-\frac{u}{8r}=0,\end{aligned}$$

where $(x,y)\in\mathbb{R}^2\backslash\{0\}$, $r=\sqrt{x^2+y^2}$, being written in polar coordinates

$$\begin{aligned}&r^2(1+r)\frac{\partial^3u}{\partial r^3}+\frac{\partial^3u}{\partial r\partial\theta^2}+3\left(r+\frac{1}{2}\right)\frac{\partial^2u}{\partial r^2}\\&\qquad-\frac{1}{2r}\frac{\partial^2u}{\partial\theta^2}+\left(r(1-\mu^2)+\frac{1}{4}\right)\frac{\partial u}{\partial r}-\frac{u}{8r}=0,\end{aligned}$$

has a solution in terms of the index integral

$$\begin{aligned}u(r,\theta)=\sqrt{\pi}\int_0^\infty&\Gamma\left(\frac{1}{2}+i\tau-\mu\right)\Gamma\left(\frac{1}{2}-i\tau-\mu\right)\\&\times\left[P^{\mu}_{-1/2+i\tau}\left(\sqrt{\frac{x+1}{x}}\right)\right]^2\frac{\sinh(\theta\tau)}{\sinh(\alpha\tau)}g(\tau)\,d\tau.\end{aligned}$$

under boundary conditions (6), where $\mathrm{Re}(\mu)<0$ and $g(\tau)$ is defined by reciprocal index transform (see details in [7])

$$g(\tau) = \frac{e^{i\mu\pi}}{\pi\sqrt{\pi}}\,\tau\sin(\pi\tau)\int_0^\infty \left(y^{-1/2}\frac{d}{dy}y^{-1/2}\right)$$
$$\times \left[P^{\mu}_{-1/2+i\tau}\left(\sqrt{\frac{y+1}{y}}\right)\left[Q^{-\mu}_{-1/2+i\tau}\left(\sqrt{\frac{y+1}{y}}\right) + Q^{-\mu}_{-1/2-i\tau}\left(\sqrt{\frac{y+1}{y}}\right)\right]\right] f(y)dy,$$

where $Q^{\mu}_{\nu}(z)$ is the associated Legendre function of the second kind.

A fourth-order PDE can be generated by the index kernel with the squares of the Kelvin functions

$$\mathrm{ker}^2_{2i\tau}(2(4x)^{1/4}) + \mathrm{kei}^2_{2i\tau}(2(4x)^{1/4}). \tag{7}$$

Indeed, in [9], the following fourth-order differential equation for the kernel (7) is established

$$x^3\frac{d^4u}{dx^4} + \frac{11}{2}x^2\frac{d^3u}{dx^3} + x\left(\frac{11}{2}+\tau^2\right)\frac{d^2u}{dx^2} + \frac{1}{2}\left(1+\tau^2\right)\frac{du}{dx} - u = 0. \tag{8}$$

Therefore, considering a boundary value problem (6) for the fourth-order PDE

$$x^2\frac{\partial^4 u}{\partial x^4} + y^2\frac{\partial^4 u}{\partial y^4} + (x^2+y^2)\frac{\partial^4 u}{\partial x^2\partial y^2} + 2xy\left(\frac{\partial^4 u}{\partial x^3\partial y} + \frac{\partial^4 u}{\partial x\partial y^3}\right)$$
$$+\frac{9}{2}\left(x\frac{\partial^3 u}{\partial x^3} + y\frac{\partial^3 u}{\partial y^3} + x\frac{\partial^3 u}{\partial x\partial y^2} + y\frac{\partial^3 u}{\partial x^2\partial y}\right)$$
$$+3\left(\frac{\partial^2 u}{\partial x^2} + \frac{\partial^2 u}{\partial y^2}\right) - \frac{u}{\sqrt{x^2+y^2}} = 0,$$

where $(x, y) \in \mathbb{R}^2\backslash\{0\}$, we write it in the polar coordinates

$$r^4\frac{\partial^4 u}{\partial r^4} + r\frac{\partial^4 u}{\partial r^2\partial\theta^2} + \frac{11}{2}r^2\frac{\partial^3 u}{\partial r^3} + \frac{1}{2}\frac{\partial^3 u}{\partial r\partial\theta^2} + \frac{11}{2}r\frac{\partial^2 u}{\partial r^2} + \frac{1}{2}\frac{\partial u}{\partial r} - u = 0.$$

Consequently, comparing with (8), the solution for the boundary value problem has the form

$$u(r,\theta) = \int_0^\infty \left[\mathrm{ker}^2_{2i\tau}(2(4r)^{1/4}) + \mathrm{kei}^2_{2i\tau}(2(4r)^{1/4})\right]\frac{\sinh(\theta\tau)}{\sinh(\alpha\tau)}\,g(\tau)\,d\tau,$$

where $g(\tau)$ is defined by reciprocal index transform, involving Kelvin functions ber and bei:

$$g(\tau) = -\frac{16\tau}{\pi}\int_0^\infty \mathrm{Im}\left[\Gamma(1+2ix)\left[\mathrm{ber}^2_{2i\tau}(2(4x)^{1/4}) + \mathrm{bei}^2_{2i\tau}(2(4x)^{1/4})\right]\right] df(x).$$

Other examples can be found in the references below.

References

1. Lebedev, N.N.: Special functions and their applications. Englewood Cliffs, N.J. : Prentice-Hall, INC (1965)
2. Yakubovich, S.: On the generalized Lebedev index transform. J. Math. Anal. Appl **429**, 184–203 (2015)
3. Yakubovich, S.: New index transforms with the product of Bessel functions. Integr. Transforms Special Funct. **26**(12), 939–955 (2015)
4. Yakubovich, S.: New index transforms of the Lebedev-Skalskaya type. Integr. Transforms and Special Funct. **27**(2), 137–152 (2016)
5. Yakubovich, S.: Lebedev's type index transforms with the modified Bessel functions. Commun. Math. Anal. **19**(2), 68–81 (2016)
6. Yakubovich, S.: Index transforms with the squares of Bessel functions. Integ. Transforms Special Funct. **27**(12), 981–994 (2016)
7. Yakubovich, S.: Lebedev type index transforms with the squares of the associated Legendre functions. Acta. Math. Hungar. **153**(1), 57–74 (2017)
8. Yakubovich, S.: Index transforms with Weber-type kernels. Integr. Transforms Special Funct. **29**(3), 171–188 (2018)
9. Yakubovich, S.: Index transforms with the squares of Kelvin functions. Integr. Transforms Special Funct. **30**(3), 190–204 (2019)
10. Yakubovich, S.: Index transforms with the product of the associated Legendre functions. Integr. Transforms Special Funct. **30**(9), 693–710 (2019)
11. Yakubovich, S.: Index transforms. World Scientific Publishing Company, Singapore (1996)
12. Zayed, A.I.: Function and Generalized Function Transformations. CRC Press, Boca Raton, New York, London, Tokyo (1996)

Summer School Short Talks

On the Blowing-Up Solutions of a System of Fractional Differential Equations

Sofwah Ahmad and Mokhtar Kirane

Abstract In this paper, we are concerned with the study of the system of fractional differential equations

$$
\begin{aligned}
{}^{c}D^{\alpha}_{0+}X(t) &= \Gamma(\alpha)(t+1)^{k_1}X^{a}(t)Y^{q}(t), \quad 0<\alpha<1, \quad t>0,\\
{}^{c}D^{\beta}_{0+}Y(t) &= \Gamma(\beta)(t+1)^{k_2}Y^{b}(t)X^{p}(t), \quad 0<\beta<1, \quad t>0,
\end{aligned}
$$

subject to

$$X(0)=X_0>0, \quad Y(0)=Y_0>0,$$

where $\Gamma(\sigma)$ stands for the Gamma function, ${}^{c}D^{\alpha}_{0+}$ stands for the Caputo fractional derivative, and a, b, p, q, k_1, and k_2 are real numbers that will be specified later. We present sufficient conditions for the non-existence of global solutions. Furthermore, we present the asymptotic growth of blowing-up solutions near the blow-up time.

Keywords Fractional derivative · Non-existence of global solutions · Asymptotic growth of blowing up solutions · Blow-up time

MSC2020: 26A33 · 34A08 · 34A34

1 Introduction

We are interested in analysing the blowing-up solutions of the following system of fractional differential equations

$$
\begin{cases}
{}^{c}D^{\alpha}_{0}X(t) &= \Gamma(\alpha)(t+1)^{k_1}X^{a}(t)Y^{q}(t), \quad 0<\alpha<1, \quad t>0,\\
{}^{c}D^{\beta}_{0}Y(t) &= \Gamma(\beta)(t+1)^{k_2}Y^{b}(t)X^{p}(t), \quad 0<\beta<1, \quad t>0,
\end{cases}
\tag{1}
$$

S. Ahmad (✉) · M. Kirane
Khalifa University of Science and Technology, Abu Dhabi, UAE
e-mail: 100059797@ku.ac.ae

M. Chatzakou et al. (eds.), *Direct and Inverse Problems with Applications*,
Trends in Mathematics 12, https://doi.org/10.1007/978-3-031-98645-1_16

subject to the initial conditions

$$X(0) = X_0 > 0, \quad Y(0) = Y_0 > 0, \tag{2}$$

where $a, b, p, q, > 1$, $\Gamma(\sigma)$ is the usual Gamma function

$$\Gamma(\sigma) = \int_0^\infty x^{\sigma-1} e^{-x}\, dx,$$

and the derivative operator under consideration is the Caputo fractional derivative [3]

$${}^cD_0^\alpha f(t) := \frac{1}{\Gamma(1-\alpha)} \int_0^t (t-\tau)^{-\alpha} f'(\tau) d\tau, \quad 0 \le t \le T.$$

Without loss of generality, we assume that $0 < \alpha \le \beta < 1$.

The problem (1) with constants $a = b = k_1 = k_2 = 0$ and $\alpha, \beta \in (0, 1)$ has been studied by Kirane and Furati [2]; they provide necessary conditions for the existence of the solution based on a suitable test function in the weak formulation of the problem. The analysis has been completed by Kirane and Malik [4]; they have obtained a bilateral estimate of the blow-up time and the profile of the blowing-up solutions. Their method is based on the nonlinear capacity method of Pohozaev [7]. Kirk, Olmstead and Roberts [5] also studied the system of the corresponding Volterra integral system as an extension of the work of Kirane-Malik [4].

2 Preliminary Results in the Classical Setting

We shall need the results of system (1) with $\alpha = \beta = 1$ for further analysis.

Lemma 1 *If $a > 1$ and $b > 1$, the solution of the system*

$$\begin{cases} \tilde{u}'(t) &= \tilde{u}^a(t)\tilde{v}_0^q, \quad t > 0, \\ \tilde{v}'(t) &= \tilde{v}^b(t)\tilde{u}_0^p, \quad t > 0, \end{cases}$$

subject to the initial conditions $\tilde{u}(0) = \tilde{u}_0 > 0$, $\tilde{v}(0) = \tilde{v}_0 > 0$, blows up in finite time.

Lemma 2 *If $p > 1$ and $q > 1$, the solution of the system*

$$\begin{cases} \tilde{u}'(t) &= \tilde{u}_0^a\tilde{v}^q(t), \quad t > 0, \\ \tilde{v}'(t) &= \tilde{v}_0^b\tilde{u}^p(t), \quad t > 0, \end{cases}$$

subject to the initial conditions $\tilde{u}(0) = \tilde{u}_0 > 0$, $\tilde{v}(0) = \tilde{v}_0 > 0$, blows up in finite time.

For subsequent discussion, let

$$\lambda = p + 1 - a \neq 0, \qquad \text{and} \qquad \mu = q + 1 - b \neq 0,$$

and consider the following conditions

$$(I) \quad \begin{cases} \lambda > 0, \ \mu > 0, \ \text{or} \\ \lambda < 0, \ \mu < 0 \ \text{ and } \ \frac{X_0^\lambda}{\lambda} = \frac{Y_0^\mu}{\mu}. \end{cases}$$

Lemma 3 *If one of the conditions of (I) is satisfied, then the solution of the system*

$$\begin{cases} \tilde{u}'(t) &= \tilde{u}^a(t)\tilde{v}^q(t), \quad t > 0, \\ \tilde{v}'(t) &= \tilde{v}^b(t)\tilde{u}^p(t), \quad t > 0, \end{cases}$$

subject to the initial conditions $\tilde{u}(0) = \tilde{u}_0 > 0$, $\tilde{v}(0) = \tilde{v}_0 > 0$, *blows up in finite time.*

3 Local Existence

We establish the existence of a pair of positive solutions to the problem (1). We consider the system (1) in its equivalent integral form

$$\begin{cases} X(t) = X_0 + \displaystyle\int_0^t (t-s)^{\alpha-1}\Big((s+1)^{k_1} X^a(s) Y^q(s)\Big)\, ds, \\ Y(t) = Y_0 + \displaystyle\int_0^t (t-s)^{\beta-1}\Big((s+1)^{k_2} Y^b(s) X^p(s)\Big)\, ds. \end{cases} \tag{3}$$

Let $T > 0$. We consider the following Banach space of continuous functions $\mathcal{K}_T = (C[0,T])^2$ equipped with the Chebyshev norm

$$\|(X, Y)\| = \|X\|_\infty + \|Y\|_\infty,$$

where $\|\cdot\|_\infty$ denotes the usual supremum norm: $\|x\|_\infty = \sup_{0\leq t\leq T} |x(t)|$.

Definition 1 *(Mild Solution)* Let $X_0, Y_0 \in \mathbb{R}^+$ and $T > 0$. We say that the pair of functions $U = (X, Y) \in \mathcal{K}_T$ is a mild solution of the system (1) if X and Y satisfy the equivalent integral system (3).

Theorem 1 (Local Existence) *Let* $X_0, Y_0 > 0$, *then there exist a* $T > 0$ *and a unique mild solution* $(X, Y) \in \mathcal{K}_T$ *to* (1).

Next, we define the maximal interval of existence $[0, T_{\max})$:

$$T_{\max} := \sup\{T > 0;\ \text{there exist a mild solution } U \in \mathcal{K}_T \text{ of}(16.3)\}$$

We have the following alternatives:

- $T_{\max} = +\infty$, or ;
- $T_{\max} < +\infty$ and $\|U\|_\infty = \|X\|_\infty + \|Y\|_\infty \to +\infty$ as $t \to T_{\max}$.

If $T_{\max} = +\infty$ we say that the solution is global.

Using the appropriate weakly singular Grönwall's inequality [6], one may establish the uniqueness in the whole interval of existence given by the following theorem.

Theorem 2 (Uniqueness) *The solution of the system* (3) *corresponding to the initial data* (2) *is unique on the interval* $[0, T]$ *for any* $T > 0$ *in the maximal interval time of existence.*

The next part of our result is to show that the latter part of the alternative holds, that is $T_{\max} < +\infty$ and $\|U\|_\infty = \|X\|_\infty + \|Y\|_\infty \to +\infty$ as $t \to T_{\max}$.

4 Blowing-Up Behaviour

In this section, we give sufficient conditions for the blow-up behaviour of the system (1)–(2).

Theorem 3 *If*

(i) $k_1 = (1-\beta)q + (1-\alpha)a$ *and* $k_2 = (1-\beta)b + (1-\alpha)p$; *or*
(ii) $k_1 = k_2 = k$ *and* $(1-\beta)q + (1-\alpha)a = (1-\beta)b + (1-\alpha)p = m$

and one of the conditions of (I) is satisfied, then the solution $(X(t), Y(t))$ *of the system* (1) *blows up at a finite time.*

Theorem 4 *Let* $k_1 = k_2 = 0$. *The solution* $(X(t), Y(t))$ *of system* (1) *blows up in finite time.*

Here we shall give the sketch of proof of Theorem 3 part (i). The proof of the other part of Theorems 3 and 4 are similar. We refer the reader to [9] for detailed proofs.

Proof (Sketch of proof of Theorem (1) part i.) Let

$$u(t) = (t+1)^{1-\alpha} X(t), \qquad \text{and} \qquad v(t) = (t+1)^{1-\beta} Y(t),$$

we obtain

$$\begin{cases} u(t) \geq \displaystyle\int_0^t \frac{(s+1)^{k_1}}{(s+1)^{(1-\alpha)a+(1-\beta)q}} u^a(s)v^q(s)\,ds, \\ v(t) \geq \displaystyle\int_0^t \frac{(s+1)^{k_2}}{(s+1)^{(1-\alpha)p+(1-\beta)b}} u^p(s)v^b(s)\,ds. \end{cases}$$

It is clear that, by differentiation and using the assumption on k_1 and k_2, $\big(u(t), v(t)\big)$ is an upper-solution for the system

$$\begin{cases} \tilde{u}' &= \tilde{u}^a(t)\tilde{v}^q(t), \quad \tilde{u}(0) = X_0, \\ \tilde{v}' &= \tilde{v}^b(t)\tilde{u}^p(t), \quad \tilde{v}(0) = Y_0, \end{cases}$$

which blows up at a finite time.

4.1 Blow-Up Time Estimate

Using the comparison method, we have the following upper blow-up time estimate.

Theorem 5 *Under the hypotheses of Theorems 3 and 4, the upper bound estimate for the blow-up time $T_{\max}$ is given by* $\min\{T^{**}_{\tilde{u}_i}, T^{**}_{\tilde{v}_i}\}$, $i = 1, 2, 3$ *with*

$$T^{**}_{\tilde{u}_1} = \frac{X_0^{1-a}}{(a-1)\,\tilde{v}_0^q}, \qquad T^{**}_{\tilde{v}_1} = \frac{Y_0^{1-b}}{(b-1)\,\tilde{u}_0^p},$$

$$T^{**}_{\tilde{u}_2} = \frac{X_0^{1-s_1-a}}{(s_1-1)} h^{\frac{-q}{q+1}}, \qquad T^{**}_{\tilde{v}_2} = \frac{Y_0^{1-s_2-b}}{(s_2-1)} h^{\frac{p}{p+1}},$$

$$T^{**}_{\tilde{u}_3} = \frac{X_0^{1-(a+\lambda q/\mu)}}{\left(a+\frac{\lambda q}{\mu}-1\right)} \left(\frac{\lambda}{\mu}\right)^{\frac{q}{\mu}}, \quad T^{**}_{\tilde{v}_3} = \frac{Y_0^{1-(b+\frac{\mu p}{\lambda})}}{\left(b+\frac{\mu p}{\lambda}-1\right)} \left(\frac{\mu}{\lambda}\right)^{\frac{p}{\lambda}},$$

where

$$h = \frac{(q+1)\tilde{v}_0^b}{(p+1)\tilde{u}_0^a}, \quad s_1 = (p+1)\frac{q}{q+1}, \quad s_2 = (q+1)\frac{p}{p+1}.$$

4.2 Behaviour of the Blowing-Up Solutions Near the Critical Time

We see from the previous subsection that there is a critical time $T_{\max}$, such that

$$X(t) \to +\infty, \qquad Y(t) \to +\infty, \qquad \text{as} \quad t \to T_{\max}.$$

Using the methods established in [1] and [8] we may specify the growth rate of the solution near the maximal time of existence $T_{\max}$ ($t \to T_{\max}$), that is given by

$$X \sim \left(\frac{\tilde{r}_1^{(b-1)}}{\tilde{r}_2^{q}}\right)^{\frac{1}{pq-(a-1)(b-1)}} \left(\frac{1}{T_{\max}-t}\right)^{\frac{\beta q-\alpha(b-1)}{pq-(a-1)(b-1)}},$$

$$Y \sim \left(\frac{\tilde{r}_2^{(a-1)}}{\tilde{r}_1^{p}}\right)^{\frac{1}{pq-(a-1)(b-1)}} \left(\frac{1}{T_{\max}-t}\right)^{\frac{\alpha p-\beta(a-1)}{pq-(a-1)(b-1)}},$$

with $\tilde{r}_1$ and $\tilde{r}_2$ given by

$$\tilde{r}_1 = r_1(T_{\max})\Gamma(\alpha)\,\frac{\Gamma\left(\frac{\beta q-\alpha(b-1)}{pq-(a-1)(b-1)}\right)}{\Gamma\left(\frac{\beta q+\alpha(pq-a(b-1))}{pq-(a-1)(b-1)}\right)},$$

$$\tilde{r}_2 = r_2(T_{\max})\Gamma(\beta)\,\frac{\Gamma\left(\frac{\alpha p-\beta(a-1)}{pq-(a-1)(b-1)}\right)}{\Gamma\left(\frac{\alpha p+\beta(pq-b(a-1))}{pq-(a-1)(b-1)}\right)}.$$

provided that $p > a - 1$ and $q > b - 1$.

Acknowledgements The first author is supported by Khalifa University through Doctorate Research and Teaching Scholarship. She would like to also extend her gratitude to Prof. Michael Ruzhansky and his research team at the Ghent Analysis & PDE Center, for their hospitality throughout her stay in Ghent.

References

1. Bleistein, N., Handelsman, R.A.: Asymptotic expansions of integrals, 2nd edn. Dover Publications Inc, New York (1986)
2. Furati, K.M., Kirane, M.: Necessary conditions for the existence of global solutions to systems of fractional differential equations. Fract. Calc. Appl. Anal. **11**, 281–298 (2008)
3. Kilbas, A.A., Srivastava, H.M., Trujillo, J.J.: Theory and applications of fractional differential Equations, North-Holland Math. Stud., 204, Elsevier Science B.V., Amsterdam, 2006
4. Kirane, M., Malik, S.A.: The profile of blowing-up solutions to a nonlinear system of fractional differential equations. Nonlinear Analysis: Theory, Methods & Applications **73**, 3723–3736 (2010)
5. Kirk, C.M., Olmstead, W.E., Roberts, C.A.: A system of nonlinear Volterra equations with blow-up solutions. J. Integr. Equ. Appl. **25**, 377–393 (2013)
6. Medved, M.: A new approach to an analysis of Henry type integral inequalities and their Bihari type versions. J. Math. Anal. Appl. **214**, 349–366 (1997)
7. Mitidieri, E.L., Pohozaev, S.I.: A priori estimates and blow-up of solutions to nonlinear partial differential equations and inequalities. Proceed. Steklov Inst. Math. **234**, 1–362 (2001)

8. Roberts, C.A., Olmstead, W.E.: Growth rates for blow-up solutions of nonlinear volterra equations. Quart. Appl. Math. **54**, 153–159 (1996)
9. Ahmad, S., Kirane, M.: On the blowing up solutions of a system of fractional differential equations. Discr. Continuous Dyn. Syst.-S **18**(2), 405–425 (2025). https://doi.org/10.3934/dcdss.2024081

O'Neil-type Inequality for Convolutions in Morrey Spaces

Diana Chigambayeva

Abstract We study the boundedness of the convolution operator in the case of Morrey spaces. Moreover, their embedding properties and relationship with the Lebesgue spaces are investigated, and basic functional inequalities for these spaces are described. Also, we consider the local Morrey-type spaces and their relationship with Morrey spaces. Additionally, their interpolation properties are described. In particular, by applying the Marcinkiewicz-type interpolation theorem, we obtain an estimate for the norm of the convolution operator in the Morrey spaces, namely the O'Neil-type inequality.

Keywords Morrey spaces · Convolution operator · Interpolation theorem · Marcinkiewicz-type interpolation theorem · O'Neil-type ineguality

Classifications 46B70 · 47A30 · 47A57 · 47A63

1 Introduction

In the modern theory of interpolation, as one of the main sections of functional analysis, Morrey spaces and their various generalizations are dynamically studied and also find great application in the theory of partial differential equations. Let us recall that classical Morrey spaces were introduced by C. Morrey in 1938 [12] when studying problems related to partial differential equations that are quasilinear elliptic. It is also important that Morrey spaces were very conveniently used to solve complex nonlinear equations of mathematical physics, such as the Euler and Navier-Stokes equations. It should also be noted that within the framework of the modern theory of functional spaces interpolation, a new dynamic and fairly confident direction of the Morrey spaces theory was formed.

D. Chigambayeva (✉)
Astana IT University, Astana, Kazakhstan
e-mail: chigambayeva.d@gmail.com

M. Chatzakou et al. (eds.), *Direct and Inverse Problems with Applications*, Trends in Mathematics 12, https://doi.org/10.1007/978-3-031-98645-1_17

Definition 1 Let $0 < p \leq \infty$ and $0 \leq \lambda \leq \frac{n}{p}$, $B_t(x)$ be the open ball with radius $t > 0$ and center at the point $x \in \mathbb{R}^n$, then $f \in M_p^\lambda$ if $f \in L_p^{loc}(\mathbb{R}^n)$, and the following norm is finite

$$\|f\|_{M_p^\lambda} \equiv ||f||_{M_p^\lambda(\mathbb{R}^n)} = \sup_{x \in \mathbb{R}^n} \sup_{t>0} t^{-\lambda} \|f\|_{L_p(B_t(x))} < \infty. \tag{1}$$

If $\lambda = 0$, then $M_p^0(\mathbb{R}^n) = L_p(\mathbb{R}^n)$, if $\lambda = \frac{n}{p}$, then $M_p^{n/p}(\mathbb{R}^n) = L_\infty(\mathbb{R}^n)$. If $\lambda < 0$ or $\lambda > \frac{n}{p}$, then $M_p^\lambda = \Theta$, where Θ is the set of all functions that are equivalent to zero on $\mathbb{R}^n$.

Let us briefly describe the properties of Morrey spaces (1), namely functional inequalities in these spaces.

Lemma 1 (Hölder's inequality in Morrey spaces)
Let $0 < p \leq p_0, p_1 \leq \infty$, $\frac{1}{p} = \frac{1}{p_0} + \frac{1}{p_1}$, $0 \leq \lambda_0 \leq \frac{n}{p_0}$, $0 \leq \lambda_1 \leq \frac{n}{p_1}$, $\lambda = \lambda_0 + \lambda_1$. *Then*

$$||fg||_{M_p^\lambda} \leq ||f||_{M_{p_0}^{\lambda_0}} ||g||_{M_{p_1}^{\lambda_1}}$$

for all functions $f \in M_{p_0}^{\lambda_0}$ *and* $g \in M_{p_1}^{\lambda_1}$.

Lemma 2 (Minkowski's inequality in Morrey spaces)
Let $0 < p \leq \infty, 0 \leq \lambda \leq \frac{n}{p}$. *Then*

$$||f + g||_{M_p^\lambda} \leq c(||f||_{M_p^\lambda} + ||g||_{M_p^\lambda})$$

for all functions $f \in M_p^\lambda$ *and* $g \in M_p^\lambda$ *and* $c > 0$ *depends only on* p *and* q.

Lemma 3 (Embedding properties for Morrey spaces)
Let $0 < p < q \leq \infty$ *and* $\frac{n}{q} - \nu = \frac{n}{p} - \lambda$, *then*

$$M_q^\nu \hookrightarrow M_p^\lambda.$$

Example 1 Let $\mu \in \mathbb{R}$, $0 < p \leq \infty$, $0 \leq \lambda \leq \frac{n}{p}$. Then

$$|x|^\mu \in M_p^\lambda(\mathbb{R}^n) \Leftrightarrow \begin{cases} \mu = \lambda - \frac{n}{p}, \ 0 < \lambda \leq \frac{n}{p}, \ p < \infty, \\ \mu = 0, \ \lambda = 0, \ p = \infty. \end{cases}$$

More details and examples can be found in [13]. Also, there are a number of papers on Morrey spaces and their applications, see [1, 3, 11].

Definition 2 Let $0 < p, q \leq \infty$, $\lambda > 0$ if $q < \infty$, $\lambda \geq 0$ if $q = \infty$. Local Morrey-type spaces $LM_{p,q,x}^\lambda$ are defined for any fixed point $x \in \mathbb{R}^n$ as the spaces of all functions $f \in L_p^{loc}(\mathbb{R}^n)$ such that for $q < \infty$

$$||f||_{LM_{p,q,x}^\lambda} = \left(\int_0^\infty \left(t^{-\lambda} \|f\|_{L_p(B_t(x))}\right)^q \frac{dt}{t} \right)^{\frac{1}{q}} < \infty, \tag{2}$$

and for $q = \infty$

$$||f||_{LM^{\lambda}_{p,\infty,x}} = \sup_{t>0} t^{-\lambda}\|f\|_{L_p(B_t(x))} < \infty.$$

The relation to the Morrey spaces (1) follows from

$$||f||_{M^{\lambda}_p} = \sup_{x\in\mathbb{R}^n} ||f||_{LM^{\lambda}_{p,\infty,x}}.$$

The local Morrey-type spaces (2) were introduced in [4], and there are some papers on Morrey-type spaces and integral operators in these spaces. The general local Morrey-type spaces and generalized Morrey-type spaces were introduced in [5, 6], and their interpolation properties were obtained in [7].

Let us for $0 < p, \theta \leq \infty$ consider the Lorentz spaces $L_{p,\theta}(\mathbb{R}^n)$ of all measurable functions $\mathbb{R}^n$, for which

$$\|f\|_{L_{p,\theta}(\mathbb{R}^n)} = \left(\int\limits_0^{\infty} \left(t^{\frac{1}{p}} f^*(t)\right)^{\theta} \frac{dt}{t}\right)^{\frac{1}{\theta}}$$

if $\theta < \infty$ and with usual modification if $\theta = \infty$, where f^* is the nonincreasing rearrangement of function f.

Let us consider the convolution operator

$$(k * f)(x) = \int\limits_{\mathbb{R}^n} k(x-y)f(y)dy$$

for all functions f and k defined on $\mathbb{R}^n$.

O'Neil's inequality for convolution has the following form

$$\|k * f\|_{L_q(\mathbb{R}^n)} \leq c\|k\|_{L_{r,\infty}(\mathbb{R}^n)}\|f\|_{L_p(\mathbb{R}^n)},$$

where $1 < p < q < \infty$, $\frac{1}{r} = 1 + \frac{1}{q} - \frac{1}{p}$ and $c > 0$ depends only on n, p and q.

The boundedness of the convolution operator in the Lorentz spaces was proved in [14, 15], in the Lebesgue spaces with weights in [9, 17]. Some analogues for inequalities with convolution operator in the Morrey-type spaces were proved in [8]. We need to obtain O'Neil's inequality for the case of Morrey spaces. Our technique is based on the interpolation properties of the Morrey-type spaces.

2 Main Results

For proving the boundedness of the convolution operator in Morrey spaces, we use the following additional results—interpolation Theorem 1 and Lemma 4, which were proved in [7].

Theorem 1 *Let* $x \in \mathbb{R}^n$, $0 < \sigma \leq \infty$, $0 \leq \nu_0, \nu_1 < \infty$, $\nu, \nu_1 > 0$ *if* $\sigma < \infty$, $\nu_0 \neq \nu_1$, $0 \leq \lambda_0, \lambda_1 < \infty$, $\lambda_0 \neq \lambda_1$, $0 < p, q \leq \infty, 0 < \theta < 1$ *and*

$$\nu = (1-\theta)\nu_0 + \theta\nu_1,\ \lambda = (1-\theta)\lambda_0 + \theta\lambda_1.$$

Let T *be a quasi-additive operator on* $\bigcup\limits_{x\in\mathbb{R}^n} \left(LM^{\nu_0}_{p,\sigma,x} + LM^{\nu_1}_{p,\sigma,x}\right)$. *Suppose that for some* $M_1, M_2 > 0$ *the following inequalities hold*

$$\|Tf\|_{LM^{\lambda_i}_{q,\infty,x}} \leq M_i \|f\|_{LM^{\nu_i}_{p,\sigma,x}}$$

for all functions $f \in LM^{\nu_i}_{p,\sigma,x}$, $i = 0, 1$, *then*

$$\|Tf\|_{M^{\lambda}_{q}} \leq cAM_0^{1-\theta} M_1^{\theta} \|f\|_{M^{\nu}_{p}}$$

for all functions $f \in M^{\nu}_{p}$, *where* $c > 0$ *depends only on* $\nu_0, \nu_1, \lambda_0, \lambda_1, p, q, \sigma, \theta$.

Lemma 4 *Let* $x \in \mathbb{R}^n$, $1 < p < q < \infty$, $0 < \nu \leq \lambda < \frac{n}{q}$,

$$\frac{1}{\rho} = 1 + \frac{1}{q} - \frac{1}{p} - \frac{\lambda-\nu}{n} \quad \text{and} \quad \frac{1}{s} = 1 + \frac{1}{q} - \frac{1}{p} - \frac{\lambda}{n}.$$

If $f \in LM^{\nu}_{p,1,x}$, $k \in L_{\rho,\infty}(\mathbb{R}^n)$ *and*

$$M(\lambda, \nu) = \sup_{m\in\mathbb{Z}} 2^{\nu m} \|k\|_{L_{s,\infty}(D_m)} < \infty,$$

where $D_m = B_{2^{m+1}}(0) \setminus B_{2^m}(0)$. *Then the convolution* $k * f$ *exists almost everywhere on* $\mathbb{R}^n$ *and*

$$\|k * f\|_{LM^{\lambda}_{q,\infty,x}} \leq c(\|k\|_{L_{\rho,\infty}(\mathbb{R}^n)} + M(\lambda,\nu))\|f\|_{LM^{\nu}_{p,1,x}},$$

where $c > 0$ *depends only on the parameters* n, p, q, ν *and* λ.

Then, we will have the following main results for this paper.

Theorem 2 *Let* $1 < p < q < \infty$, $0 < \nu \leq \lambda < \frac{n}{q}$, $\mu > \nu$,

$$\frac{1}{\rho} = 1 + \frac{1}{q} - \frac{1}{p} - \frac{\lambda-\nu}{n} \quad \text{and} \quad \frac{1}{\sigma} = \frac{1}{\rho} - \frac{\mu}{n} > 0.$$

If $f \in M^{\nu}_{p}$, $k \in L_{\rho,\infty}(\mathbb{R}^n)$ *and*

$$N(\lambda, \mu, \nu) = \sup_{m\in\mathbb{Z}} 2^{\mu m} \|k\|_{L_{\sigma,\infty}(D_m)} < \infty,$$

then the convolution $k * f$ exists almost everywhere on $\mathbb{R}^n$ and

$$||k * f||_{M_q^\lambda} \leq c(||k||_{L_{\rho,\infty}(\mathbb{R}^n)} + N(\lambda, \mu, \nu))||f||_{M_p^\nu},$$

where $c > 0$ depends only on the parameters $n, p, q, \lambda, \mu, \nu$.

Acknowledgements The author would like to thank the Bolashak Scholars' international internship and the Ghent Analysis and PDE Center of Ghent University. Also, the author expresses her gratitude to Professor M. Ruzhansky for his valuable advices.

References

1. Adams, D.R.: Morrey spaces. Birkhäuser. 124 (2015)
2. Blozinski, A.P.: On a convolution theorem for L_{pq} spaces. Trans. Amer. Math. Soc. **164**, 255–265 (1972)
3. Burenkov, V.I.: Recent progress in studing the boundedness of classical operators of real analysis in general Morrey-type spaces I. Eurasian Math. J. **3**(3), 11–32 (2012)
4. Burenkov, V.I., Guliyev, H.V.: Nessesary and sufficient conditions for the boundedness of the maximal operator in the local Morrey-type spaces. Doklady Ross. Akad. Nauk. Matematika. **391**(5), 591–594 (2003)
5. Burenkov, V.I., Nursultanov, E.D., Chigambayeva, D.K.: Description of interpolation spaces for the pair of local Morrey-type spaces and their generalizations. Proc. Steklov Inst. Math. **284**, 105–137 (2014)
6. Burenkov, V.I., Nursultanov, E.D., Chigambayeva, D.K.: Marcinkiewicz-type interpolation theorem and estimates for convolutions for Morrey-type spaces. Eurasian Math. J. **9**(2), 82–88 (2018)
7. Burenkov, V.I., Nursultanov, E.D., Chigambayeva, D.K.: Marcinkiewicz-type interpolation theorem for Morrey-type spaces and its corollaries. Complex Var. Elliptic Equ. **65**(1), 87–108 (2020)
8. Burenkov, V.I., Tararykova, T.V.: An analog of Young's inequality for convolutions of functions for general Morrey-type spaces. Trudy Math. Inst. Steklov **293**, 107–126 (2016)
9. Kerman, R.A.: Convolution theorems with weights. Trans. Amer. Math. Soc. **280**(1), 207–219 (1983)
10. Kostyuchenko, A.G., Nursultanov, E.D.: On integral operators in L_p-spaces. Fundam. Prikl. Mat. **5**(2), 475–491 (1999)
11. Lemarié-Rieusset, P.G.: The role of Morrey spaces in the study of Navier-Stokes and Euler equations. Eurasian Math. J. **3**(3), 62–93 (2012)
12. Morrey, C.B.: On the solution of quasi-linear elliptic partial differential equations. Trans. Amer. Math. Soc. **43**, 126–166 (1938)
13. Nursultanov, E.D., Chigambayeva, D.K.: Interpolation of Morrey-type spaces: Study Guide. Evero. 137 (2019)
14. Nursultanov, E.D., Tikhonov, S.: Convolution inequalities in Lorentz spaces. J. Fourier Anal. Appl. **17**, 486–505 (2011)

15. O'Neil, R.: Convolution operators and $L(p, q)$ spaces. Duke Math. J. **30**, 129–142 (1963)
16. Peetre, J.: On convolution operators leaving $L^{p,\lambda}$ spaces invariant. Ann. Mat. Pura ed Appl. **72**(4), 295–304 (1966)
17. Stepanov, V.D.: Some topics in the theory of integral convolution operators. Dalnauka, Vladivostok (2000)

On Direct and Inverse Problems for Mixed Wave-Diffusion Equation with the Riemann-Liouville Fractional Derivative

Erkinjon Karimov and Maftuna Mirzaeva

Abstract Direct and inverse problem of finding unknown time-moment for mixed wave-diffusion equation involving the Riemann-Liuoville fractional derivative have been targeted for a unique solvability. A solution of a first boundary problem for the time-fractional diffusion-wave equation, obtained using Green's function, played a key role during evaluations. Certain properties of the Wright-type function were effectively utilized, as well.

Keywords Time-fractional wave-diffusion equation · Wright-type function · Green's function · Inverse problem

MSC2020: 35M13 · 35R11 · 35R30

1 Introduction

It is well-known that the wave and diffusion processes are modeled by the wave (hyperbolic type) equation and diffusion (parabolic type) equation. Due to the rapid development of Fractional Calculus, such processes started to be modeled using the fractional analogues (mathematical generalizations) of wave and diffusion equations [1]. Appropriate motivations were described based on mathematically universal properties of fractional derivatives, which allow considering more options during the idealization of the corresponding process [2]. In numerous works, direct and inverse problems for fractional order diffusion and wave equations have been studied. Omitting such works, we would like to mention works which targeted combinations of fractional order diffusion and wave equations. Let us start with the pioneering work [3] in which the combination of sub-diffusion and degenerate hyperbolic equation

E. Karimov (✉)
Ghent University, Ghent, Belgium
e-mail: erkinjon.karimov@ugent.be

M. Mirzaeva
Fergana State University, Fergana, Uzbekistan

M. Chatzakou et al. (eds.), *Direct and Inverse Problems with Applications*,
Trends in Mathematics 12, https://doi.org/10.1007/978-3-031-98645-1_18

was considered. Due to the fundamental result of Pskhu [4], in which solutions of boundary-value problems for diffusion-wave equation with the Riemann-Liouville fractional derivative were represented in an explicit form, a series of works devoted to the investigation of mixed equations involving the same sub-diffusion equation, were published (see, e.g. [5]). Another type of investigation appeared based on the method of separation of variables, which allowed one to find series-form solutions to direct and inverse problems for mixed equations involving various fractional derivatives [6–9].

We will note the recent work by Slodička [10], where variable-order fractional differential equations have been considered, which one can consider as a combination of several equations with different fractional order.

2 Direct Problem

Direct problem. To find a function $u(t,x)$, satisfying the following

- regularity conditions $u(t,x)\in\Omega$; $u(\cdot,x)\in C^1[0,T_*]\cap {}^{RL}D_{0t}^{\beta_1}u(\cdot,x)\in C(0,T_*)$; ${}^{RL}D_{T_*t}^{\beta_2}u(\cdot,x)\in C(T_*,T)$, $u(t,\cdot)\in C^2(0,1)$;
- the following equation in Ω

$$ {}^{RL}D_{a(t),t}^{\beta(t)}u(t,x)-u_{xx}(t,x)=f(t,x); \tag{1}$$

- initial condition

$$\lim_{t\to 0} I_{0t}^{2-\beta_1}u(t,x)=\tau(x),\quad 0<x<1 \tag{2}$$

or

$$\lim_{t\to 0} {}^{RL}D_{0t}^{\beta_1-1}u(t,x)=\nu(x),\quad 0<x<1; \tag{3}$$

- boundary conditions

$$u(t,0)=\varphi_1(t),\,u(t,1)=\varphi_2(t),\,0<t<T; \tag{4}$$

- transmitting conditions

$$\lim_{t\to T_*+0} {}^{RL}D_{T_*t}^{\beta_2}u(t,x)=u_t(T_*-0,x),$$

$$\lim_{t\to T_*+0} I_{T_*t}^{1-\beta_2}u(t,x)=\lim_{t\to T_*-0} t^{2-\beta_1}u(t,x),\,0<x<1. \tag{5}$$

In the sequel, we denote by

$$\Omega = \Omega_1 \cup \Omega_2 \cup J, \quad \Omega_1 = \{(t, x) : \ 0 < x < 1, \ 0 < t < T_*\},$$

$$\Omega_2 = \{(t, x) : \ 0 < x < 1, \ T_* < t < T\}, \quad J = \{(t, x) : \ 0 < x < 1, \ t = T_*\},$$

$$a(t) = \begin{cases} 0, & 0 \le t < T_*, \\ T_*, & T_* \le t < T; \end{cases} \quad \beta(t) = \begin{cases} \beta_1, & 0 \le t \le T_*, \\ \beta_2, & T_* \le t < T; \end{cases}$$

$$0 < T_* < T \in \mathbb{R}^+, \ 1 < \beta_1 \le 2, \ 0 < \beta_2 \le 1.$$

In this problem, we will use solutions of the first boundary problem for the Eq. (1) in Ω_1 and Ω_2, which was obtained using the method of Green's function [4]:

$$u(t, x) = \int_0^t \varphi_1(\eta) G_\xi(x, t, 0, \eta) d\eta - \int_0^t \varphi_2(\eta) G_\xi(x, t, 1, \eta) d\eta +$$
$$\int_0^1 \tau(\xi) G(x, t, \xi, 0) d\xi - \int_0^1 \nu(\xi) G_\eta(x, t, \xi, 0) d\xi +$$
$$\int_0^t \int_0^1 f(\xi, \eta) G(x, t, \xi, \eta) d\xi d\eta$$

in Ω_1

$$u(t, x) = \int_{T_*}^t \varphi_1(\eta) G^*_\xi(x, t, 0, \eta) d\eta - \int_{T_*}^t \varphi_2(\eta) G^*_\xi(x, t, 1, \eta) d\eta +$$
$$\int_0^1 \tau_*(\xi) G^*(x, t, \xi, T_*) d\xi + \int_{T_*}^t \int_0^1 f(\xi, \eta) G^*(x, t, \xi, \eta) d\xi d\eta$$

in Ω_2. Here

$$G(x, t, \xi, \eta) =$$
$$\frac{(t-\eta)^{\frac{\beta_1}{2}-1}}{2} \sum_{n=-\infty}^{\infty} \left[e^{1, \frac{\beta_1}{2}}_{1, \frac{\beta_1}{2}} \left(-\frac{|x-\xi+2n|}{(t-\eta)^{\frac{\beta_1}{2}}} \right) - e^{1, \frac{\beta_1}{2}}_{1, \frac{\beta_1}{2}} \left(-\frac{|x+\xi+2n|}{(t-\eta)^{\frac{\beta_1}{2}}} \right) \right],$$

$$G^*(x,t,\xi,\eta)=$$
$$\frac{(t-\eta)^{\frac{\beta_2}{2}-1}}{2}\sum_{n=-\infty}^{\infty}\left[e^{1,\frac{\beta_2}{2}}_{1,\frac{\beta_2}{2}}\left(-\frac{|x-\xi+2n|}{(t-\eta)^{\frac{\beta_2}{2}}}\right)-e^{1,\frac{\beta_2}{2}}_{1,\frac{\beta_2}{2}}\left(-\frac{|x+\xi+2n|}{(t-\eta)^{\frac{\beta_2}{2}}}\right)\right],$$

also $e^{\mu,\delta}_{\alpha,\beta}(z)=\sum\limits_{k=0}^{\infty}\frac{z^k}{\Gamma(\alpha k+\mu)\Gamma(\delta-\beta k)}$ is the Wright type function [4].

We first apply transmitting condition $\lim\limits_{t\to T_*+0}{}^{RL}D^{\beta_2}_{T_*t}u(t,x)=u_t(T_*-0,x)$ and will obtain the following equality:

$$\lim_{t\to T_*+0}\left\{-\tau'_*(1)\frac{(t-T_*)^{\frac{\beta_2}{2}-1}}{2}\sum_{n=-\infty}^{\infty}\left[e^{1,\frac{\beta_2}{2}}_{1,\frac{\beta_2}{2}}\left(-\frac{|x-1+2n|}{(t-T_*)^{\frac{\beta_2}{2}}}\right)-e^{1,\frac{\beta_2}{2}}_{1,\frac{\beta_2}{2}}\left(-\frac{|x+1+2n|}{(t-T_*)^{\frac{\beta_2}{2}}}\right)\right]+\right.$$
$$\left.\int_0^1\tau''_*(\xi)\,G^*(x,t,\xi,T_*)\,d\xi\right\}=$$
$$\varphi_1(0)\,G_\xi(x,T_*,0,0)+\int_0^{T_*}\varphi'_1(\eta)\,G_\xi(x,T_*,0,\eta)\,d\eta+$$
$$\varphi_2(0)\,G_\xi(x,T_*,1,0)+\int_0^{T_*}\varphi'_2(\eta)\,G_\xi(x,T_*,1,\eta)\,d\eta+$$
$$\int_0^1\tau(\xi)\,G_t(x,T_*,\xi,0)\,d\xi-\int_0^1\nu(\xi)\,G_{\eta t}(x,T_*,\xi,0)\,d\xi+$$
$$\int_0^{T_*}\int_0^1 f(\xi,\eta)\,G_t(x,T_*,\xi,\eta)\,d\xi d\eta.$$

Now, we use the second transmitting condition:

$$\tau_*(x)=T_*^{2-\beta_1}\cdot\left[\int_0^{T_*}\varphi_1(\eta)\,G_\xi(x,T_*,0,\eta)\,d\eta-\int_0^{T_*}\varphi_2(\eta)\,G_\xi(x,T_*,1,\eta)\,d\eta+\right.$$
$$\int_0^1\tau(\xi)\,G(x,T_*,\xi,0)\,d\xi-\int_0^1\nu(\xi)\,G_\eta(x,T_*,\xi,0)\,d\xi+$$
$$\left.\int_0^{T_*}\int_0^1 f(\xi,\eta)\,G(x,T_*,\xi,\eta)\,d\xi d\eta\right].$$

In fact, we have got the following system of equations with respect to the unknown function $\tau_*(x)$ and function $\nu(x)$ or $\tau(x)$ (it depends on which function is given in initial condition):

$$\begin{cases} \lim\limits_{t\to T_*+0}\left\{-\tau'_*(1)\dfrac{(t-T_*)^{\frac{\beta_2}{2}-1}}{2}\times\right. \\ \sum\limits_{n=-\infty}^{\infty}\left[e^{1,\frac{\beta_2}{2}}_{1,\frac{\beta_2}{2}}\left(-\dfrac{|x-1+2n|}{(t-T_*)^{\frac{\beta_2}{2}}}\right)-e^{1,\frac{\beta_2}{2}}_{1,\frac{\beta_2}{2}}\left(-\dfrac{|x+1+2n|}{(t-T_*)^{\frac{\beta_2}{2}}}\right)\right]+ \\ \left.\int\limits_0^1 \tau''_*(\xi)\,G^*(x,t,\xi,T_*)\,d\xi\right\}=h_1(x,T_*), \\ \tau_*(x)=h_2(x,T_*), \end{cases} \tag{6}$$

where

$$h_1(x,T_*)=\varphi_1(0)\,G_\xi(x,T_*,0,0)+\int\limits_0^{T_*}\varphi'_1(\eta)\,G_\xi(x,T_*,0,\eta)\,d\eta+$$

$$\varphi_2(0)\,G_\xi(x,T_*,1,0)+\int\limits_0^{T_*}\varphi'_2(\eta)\,G_\xi(x,T_*,1,\eta)\,d\eta+$$

$$\int\limits_0^1\tau(\xi)\,G_t(x,T_*,\xi,0)\,d\xi-\int\limits_0^1\nu(\xi)\,G_{\eta t}(x,T_*,\xi,0)\,d\xi+$$

$$\int\limits_0^{T_*}\int\limits_0^1 f(\xi,\eta)\,G_t(x,T_*,\xi,\eta)\,d\xi d\eta$$

and

$$h_2(x,T_*)=T_*^{\,2-\beta_1}\left[\int\limits_0^{T_*}\varphi_1(\eta)\,G_\xi(x,T_*,0,\eta)\,d\eta-\int\limits_0^{T_*}\varphi_2(\eta)\,G_\xi(x,T_*,1,\eta)\,d\eta+\right.$$

$$\int\limits_0^1\tau(\xi)\,G(x,T_*,\xi,0)\,d\xi-\int\limits_0^1\nu(\xi)\,G_\eta(x,T_*,\xi,0)\,d\xi+$$

$$\left.\int\limits_0^{T_*}\int\limits_0^1 f(\xi,\eta)\,G(x,T_*,\xi,\eta)\,d\xi d\eta\right].$$

Using the second equation of the system (6), we find the first and second-order derivatives of $\tau_*(x)$ and put them in the first equation of the system reducing it to the integral equation with respect to other unknown function.

3 Inverse Problem

Inverse problem. To find a function $u(t,x)$ and unknown time-moment T_*, satisfying

- regularity conditions $u(t,x)\in\Omega$; $u(\cdot,x)\in C^1[0,T_*]\cap {}^{RL}D_{0t}^{\beta_1}u(\cdot,x)\in C(0,T_*)$; ${}^{RL}D_{T_*t}^{\beta_2}u(\cdot,x)\in C(T_*,T)$, $u(t,\cdot)\in C^2(0,1)$;
- equation (1) in Ω;
- initial conditions (2)–(3);
- boundary conditions (4);
- transmitting conditions (5).

The inverse problem will be investigated similarly, and the problem will be reduced to the system of Eq. (6). Finally, with respect to unknown time-moment T_* one can obtain the following transcendental equation:

$$\lim_{t\to T_*+0}\left\{-\left.\frac{\partial h_2(x,T_*)}{\partial x}\right|_{x=1}\cdot\frac{(t-T_*)^{\frac{\beta_2}{2}-1}}{2}\times\right.$$

$$\sum_{n=-\infty}^{\infty}\left[e^{1,\frac{\beta_2}{2}}_{1,\frac{\beta_2}{2}}\left(-\frac{|x-1+2n|}{(t-T_*)^{\frac{\beta_2}{2}}}\right)-e^{1,\frac{\beta_2}{2}}_{1,\frac{\beta_2}{2}}\left(-\frac{|x+1+2n|}{(t-T_*)^{\frac{\beta_2}{2}}}\right)\right]+$$

$$\left.+\int_0^1\frac{\partial^2 h_2(\xi,T_*)}{\partial\xi^2}G^*(x,t,\xi,T_*)\,d\xi\right\}=h_1(x,T_*),$$

where

$$\frac{\partial h_2(x,T_*)}{\partial x}=T_*^{\,2-\beta_1}\cdot\left[\int_0^{T_*}\varphi_1(\eta)\frac{\partial G_\xi(x,T_*,0,\eta)}{\partial x}d\eta-\int_0^{T_*}\varphi_2(\eta)\frac{\partial G_\xi(x,T_*,1,\eta)}{\partial x}d\eta+\right.$$

$$+\int_0^1\tau(\xi)\frac{\partial G(x,T_*,\xi,0)}{\partial x}d\xi-\int_0^1\nu(\xi)\frac{\partial G_\eta(x,T_*,\xi,0)}{\partial x}d\xi+$$

$$\left.+\int_0^{T_*}\int_0^1 f(\xi,\eta)\frac{\partial G(x,T_*,\xi,\eta)}{\partial x}d\xi d\eta\right].$$

$$\frac{\partial^2 h_2(x, T_*)}{\partial x^2} =$$

$$= T_*^{2-\beta_1} \cdot \left[\int_0^{T_*} \varphi_1(\eta) \frac{\partial^2 G_\xi(x, T_*, 0, \eta)}{\partial x^2} d\eta - \int_0^{T_*} \varphi_2(\eta) \frac{\partial^2 G_\xi(x, T_*, 1, \eta)}{\partial x^2} d\eta + \right.$$

$$+ \int_0^1 \tau(\xi) \frac{\partial^2 G(x, T_*, \xi, 0)}{\partial x^2} d\xi - \int_0^1 \nu(\xi) \frac{\partial^2 G_\eta(x, T_*, \xi, 0)}{\partial x^2} d\xi +$$

$$\left. + \int_0^{T_*} \int_0^1 f(\xi, \eta) \frac{\partial^2 G(x, T_*, \xi, \eta)}{\partial x^2} d\xi d\eta \right].$$

We note that the full details of proof can be found in recently published work [11].

4 Conclusion

In this short note, we formulate the direct and inverse problems for the time-fractional mixed equation, combining the sub-diffusion and fractional order wave equation in a rectangular domain. In [12], the similar inverse problem of finding an unknown time-moment was investigated. The distinctive side of this research is the following:

- Instead of a regularized fractional derivative (Caputo), we have used the Riemann-Liouville fractional derivative;
- Instead of using the method of separation of variables, we have used the method of Green's function;
- Obtained transcendental equation is more simple.

Acknowledgements Erkinjon Karimov is supported by the Methusalem program of the Ghent University Special Research Fund (BOF) (Grant number 01M01021).

References

1. Uchaikin, V.V.: Fractional derivatives for physicists and engineers. Volume I: Applications. Springer, Singapore (2025)
2. Povstenko, Y.: Linear fractional diffusion-wave equation for scientists and engineers. Birkhäuser (2015)
3. Kilbas, A.A., Repin, O.A.: An analog of the Tricomi problem for a mixed type equation with a partial fractional derivative. Fract. Calculus Appl. Anal. **13**(1), 69–84 (2010)

4. Pskhu, A.V.: Solution of boundary value problems for a diffusion equation of fractional order by the Green's function method. Differ. Equ. **39**(10), 1509–1513 (2003)
5. Kerbal, S., Karimov, E., Rakhmatullaeva, N.: Non-local boundary problem with integral form transmitting condition for fractional mixed type equation in a composite domain. Math. Modell. Nat. Phenomena **12**(3), 95–104 (2017)
6. Karimov, E.T., Kerbal, S., Al-Salti, N.: Inverse source problem for multi-term fractional mixed type equation. In: Ruzhansky, M., Cho, Y., Agarwal, P., Area, I. (eds) Advances in Real and Complex Analysis with Applications. Trends in Mathematics (2017). Birkhäuser, Singapore. https://doi.org/10.1007/978-981-10-4337-6_13
7. Karimov, E., Tokmagambetov, N., Khasanov, S.: Inverse problems for time-fractional mixed equation involving the caputo fractional derivative. In: Ruzhansky, M., Van Bockstal, K. (eds) Extended Abstracts 2021/2022. APDEGS 2021. Trends in Mathematics (2024), vol 2. Birkhäuser, Cham.https://doi.org/10.1007/978-3-031-42539-4_20
8. Karimov, E., Tokmagambetov, N., Toshpulatov, M.: On a mixed equation involving prabhakar fractional order integral-differential operators. In: Ruzhansky, M., Van Bockstal, K. (eds) Extended Abstracts 2021/2022. APDEGS 2021. Trends in Mathematics(2024), vol 2. Birkhäuser, Cham.https://doi.org/10.1007/978-3-031-42539-4_25
9. Karimov, E., Sobirov, Z., Khujakulov, J.: A non-local problem for a time-fractional differential equation with the Hilfer operator on metric graph. Math. Meth. Appl. Sci. **47**(18), 13562–13580 (2024)
10. Slodička, M.: A semilinear diffusion PDE with variable order time-fractional Caputo derivative subject to homogeneous Dirichlet boundary conditions. Fract. Calc. Appl. Anal. **28**, 38–75 (2025). https://doi.org/10.1007/s13540-024-00352-7
11. Karimov, E., Mirzaeva, M.: Unknown time-moment inverse problem for time-fractional mixed diffusion-wave equation. Bull. Inst. Math. **7**(6), 45–55 (2024)
12. Karimov, E.T., Tokmagambetov, N.E.: Inverse problem on finding unknown time-moment for mixed wave-diffusion equation. Lobachevskii J. Math. **45**(7), 3314–3322 (2024)

A Heuristic Rule for Landweber Iteration in Banach Spaces

Rommel R. Real

Abstract We consider the Landweber iteration for solving linear and nonlinear inverse problems in Banach spaces. Based on the discrepancy principle, we propose a heuristic parameter choice rule for choosing the regularization parameter which does not require the information on the noise level, so it is purely data-driven. According to a famous veto, convergence in the worst-case scenario cannot be expected in general. However, by imposing certain conditions on the noisy data, we establish a new convergence result which, in addition, requires neither the Gâteaux differentiability of the forward operator nor the reflexivity of the image space. This expands the range of the applications of the Landweber iteration to non-smooth, ill-posed inverse problems and to situations where the data is contaminated by various types of noise.

Keywords Landweber iteration · Inverse problems · Hanke-raus rule

MSC2020: 65J20 · Numerical solutions of ill-posed problems in abstract spaces · regularization

1 Introduction

Consider ill-posed problems of the form

$$F(x) = y^{\dagger} \tag{1}$$

where $F : \mathcal{D}(F) \subset \mathcal{X} \to \mathcal{Y}$ is a linear or a nonlinear operator, with domain $\mathcal{D}(F)$. We assume (1) has a solution. To find a solution of (1) with specific properties, we may use a proper, lower semi-continuous, uniformly convex function $\mathcal{R} : \mathcal{X} \to (-\infty, \infty]$. Let $\partial\mathcal{R}$ denote the subdifferential of R, i.e.

R. R. Real (✉)
Department of Mathematics, Physics and Computer Science, University of the Philippines Mindanao, Mindanao, Philippines
e-mail: rrreal@up.edu.ph

M. Chatzakou et al. (eds.), *Direct and Inverse Problems with Applications*, Trends in Mathematics 12, https://doi.org/10.1007/978-3-031-98645-1_19

$$\partial\mathcal{R}(x) := \left\{\xi \in \mathcal{X}^* : \mathcal{R}(z) \geq \mathcal{R}(x) + \langle \xi, z - x\rangle_{\mathcal{X}^*,\mathcal{X}} \text{ for all } z \in \mathcal{X}\right\}$$

for all x in $\mathcal{X}$, with $\mathcal{X}^*$ denoting the dual space of a Banach space $\mathcal{X}$ and $\langle\cdot,\cdot\rangle_{\mathcal{X}^*,\mathcal{X}}$ the duality pairing between $\mathcal{X}^*$ and $\mathcal{X}$. Also we denote by $\|\cdot\|_{\mathcal{X}}$ the norm on $\mathcal{X}$. We use

$$D_\xi\mathcal{R}(z, x) := \mathcal{R}(z) - \mathcal{R}(x) - \langle \xi, z - x\rangle_{\mathcal{X}^*,\mathcal{X}}, \quad z \in \mathcal{X}$$

to denote the Bregman distance induced by $\mathcal{R}$ at x in the direction ξ. By picking $x_0 \in \mathcal{D}(\partial\mathcal{R})$ and $\xi_0 \in \partial\mathcal{R}(x_0)$ as initial guesses, we define $x^\dagger$ to be a solution of (1) with the property

$$D_{\xi_0}\mathcal{R}(x^\dagger, x_0) := \min_{x\in\mathcal{D}(F)} \left\{D_{\xi_0}\mathcal{R}(x, x_0) \;:\; F(x) = y^\dagger\right\}. \tag{2}$$

The exact data $y^\dagger$ is either inaccessible or not precisely available. Instead, a noisy measurement $y^\delta \in \mathcal{Y}$ is available that satisfies

$$\|y^\delta - y^\dagger\|_{\mathcal{Y}} \leq \delta, \tag{3}$$

where δ is the noise level. Then the Landweber iteration in Banach spaces has the form

$$\xi^\delta_{n+1} = \xi^\delta_n - \mu^\delta_n L(x^\delta_n)^* j^{\mathcal{Y}}_r\left(F(x^\delta_n) - y^\delta\right), \tag{4}$$

$$x^\delta_{n+1} = \arg\min_{x\in\mathcal{X}} \left\{\mathcal{R}(x) - \langle\xi^\delta_{n+1}, x\rangle_{\mathcal{X}^*,\mathcal{X}}\right\}, \tag{5}$$

where $\xi^\delta_0 = \xi_0$, $x^\delta_0 = x_0$, the step size is given by

$$\mu^\delta_n := \tilde{\mu}^\delta_n \left\|F(x^\delta_n) - y^\delta\right\|^{p-r}_{\mathcal{Y}}, \tag{6}$$

where

$$\tilde{\mu}^\delta_n := \min\left\{\frac{\mu_0 \left\|F(x^\delta_n) - y^\delta\right\|^{p(r-1)}_{\mathcal{Y}}}{\left\|L(x^\delta_n)^* j^{\mathcal{Y}}_r\left(F(x^\delta_n) - y^\delta\right)\right\|^p_{\mathcal{X}^*}}, \mu_1\right\}, \tag{7}$$

for some positive constants μ_0 and μ_1, p is a positive constant depending on the convexity of $\mathcal{R}$, $\{L(x) : x \in \mathcal{D}(F)\}$ is a family of bounded linear operators from $\mathcal{X}$ to $\mathcal{Y}$ satisfying certain properties, and $j^{\mathcal{Y}}_r : \mathcal{Y} \to \mathcal{Y}^*$, with $1 \leq r < \infty$ is a selection of the (possibly) multi-valued mapping $J^{\mathcal{Y}}_r : \mathcal{Y} \to 2^{\mathcal{Y}^*}$, defined as the subdifferential of the convex function $y \mapsto \|y\|^r_{\mathcal{Y}}/r$.

The iteration must be terminated appropriately to produce a useful approximate solution. The authors in [13] used the discrepancy principle i.e., the iteration (4)–(5) is terminated after $n_\delta := n(\delta, y^\delta)$ steps with

$$\left\|y^\delta - F(x^\delta_{n_\delta})\right\|_{\mathcal{Y}} \leq \tau\delta < \left\|y^\delta - F(x^\delta_n)\right\|_{\mathcal{Y}}, \quad 0 \leq n < n_\delta, \tag{8}$$

where τ is an appropriately chosen constant, and the regularization property of x_n^δ as $\delta \to 0$ has been extensively studied [3, 10, 13, 15, 19]. All previous convergence analysis of Landweber iteration (under the discrepancy principle) requires standard assumptions on the forward operator F, as well as the following assumptions.

Assumption 1 1. the Banach space $\mathcal{Y}$ is uniformly smooth and $1 < r < \infty$;
2. the mapping $x \to L(x)$ from $\mathcal{X}$ to $\mathcal{L}(\mathcal{X}, \mathcal{Y})$ is continuous,

□

where $\mathcal{L}(\mathcal{X}, \mathcal{Y})$ denotes the Banach space of all bounded linear operators from the Banach space $\mathcal{X}$ to the Banach space $\mathcal{Y}$. The paper [17] relaxes Assumption 1 for key situations where the noisy data is contaminated by non-Gaussian noise.

The discrepancy principle (8) requires knowledge of the noise level, which is not always available or reliable in real-world applications. Overestimation or underestimation of noise level may lead to a significant loss of reconstruction accuracy when using the discrepancy principle. Therefore, it is necessary to develop heuristic rules for Landweber iteration that do not use any knowledge of the noise level. Based on the discrepancy principle, we will propose a heuristic rule for Landweber iteration in the spirit of the work of Hanke and Raus [7].

Bakushinskii's veto ([2]) states that heuristic rules can not lead to convergence in the sense of worst-case scenario for any regularization methods. Despite this veto, by imposing certain conditions on the noisy data, convergence results under heuristic rules are obtained for various regularization methods [6, 9, 12, 14, 16, 20]. The Landweber iteration is computationally cheap to implement, thus it is worth analyzing under heuristic rules.

In this paper, we briefly give an overview of the convergence of the Landweber iteration in Banach spaces under the Hanke-Raus heuristic rule. To do this, we first outline the assumptions lifted from previous convergence analysis of the iteration, including a compactness condition. Then, we present the Hanke-Raus rule for the iteration and the special assumption for our analysis, the noise condition. Finally, we present the main result on the convergence analysis of the Landweber iteration under the Hanke-Raus rule.

2 Landweber Iteration in Banach Spaces

To analyze the Landweber iteration in Banach spaces defined by (4)–(6), we used the standard assumptions concerning the regularization functional $\mathcal{R}$ and the forward operator F, as also used in [13, 17]. When $\mathcal{X}$ is a reflexive Banach space, these standard assumption imply that $x^\dagger$ exists and is uniquely defined [13, Lemma 3.2].

The regularization property of Landweber iteration (4)–(6) under the discrepancy principle (8) for solving (1) is established in [13, Theorem 3.9]. However, this depends heavily on the uniform smoothness of $\mathcal{Y}$ and the continuity of the mapping $\mathcal{X} \ni x \mapsto L(x) \in \mathcal{L}(\mathcal{X}, \mathcal{Y})$. There are situations where the noisy data is contaminated by non-Gaussian noise, such as the impulsive noise or the uniform distributed noise; in such situations one may choose $\mathcal{Y}$ to be the L^1-space or the L^∞-space to effectively remove the effects of noise ([1, 4]), both which are not uniformly smooth. On the other hand, there exist ill-posed inverse problems where the forward operator F is non-smooth, i.e. not necessarily Gâteaux differentiable ([5]). The paper [17] revisits the iteration (4)–(8) to provide a new convergence result that requires neither the reflexivity of $\mathcal{Y}$ nor the continuity of the mapping $x \to L(x)$. To replace Assumption 1, the following compactness condition was used.

Assumption 2 There exists a Banach space Z such that

$$\bigcup_{x\in B_{3\rho}(x_0)} \text{Range}(L(x)^*) \subset Z$$

and Z is compactly embedded into $\mathcal{X}^*$. Moreover, there is a constant $\hat{B}$ such that $\|L(x)^*\|_{\mathcal{L}(\mathcal{Y}^*,Z)} \le \hat{B}$ for all $x \in B_{3\rho}(x_0)$. □

The performance of the discrepancy principle depends heavily on the accurate knowledge of noise level. Such noise level information, however, is not always available or reliable in applications.

Based on modifying the discrepancy principle, we propose the following heuristic rule for Landweber iteration in Banach spaces in the spirit of [7].

Rule 1 *Let $a \ge 1$ be a fixed number, and let*

$$\Theta(n, y^\delta) := (n+a) \left\| F(x_n^\delta) - y^\delta \right\|_{\mathcal{Y}}^p .$$

We define $n_* := n_*(y^\delta)$ be an integer such that

$$n_* \in \text{argmin} \left\{ \Theta(n, y^\delta) : 0 \le n \le n_\infty \right\},$$

where $n_\infty := n_\infty(y^\delta)$ is the largest integer such that $x_n^\delta \in \mathcal{D}(F)$ for all $0 \le n \le n_\infty$. □

The Hanke-Raus rule has been studied in variational regularization in Banach spaces [11], and has been generalized for variational regularization in general topological spaces [16]. Rule 1 can be easily implemented. During the iteration, the value of $\Theta(n, y^\delta)$ is recorded versus n. After performing a large number of iterations, we stop and choose n_* to be the integer that minimizes $\Theta(n, y^\delta)$. Setting a large enough upper limit in the number of iterations is also a good practice to prevent badly estimating n_* [8]. The use of a fixed number a in rule 1 is a scheme first introduced in [12] to prevent the regularization parameter n_* from becoming too small.

With the integer $n_* := n_*(y^\delta)$ determined by Rule 1, we will use $x^\delta_{n_*(y^\delta)}$ as an approximate solution. A natural question is, for a family of noisy data $\{y^\delta\}$ with $y^\delta \to y$ as $\delta \to 0$, if it is possible to guarantee the convergence of $x^\delta_{n_*(y^\delta)}$ to a solution of (1) as $\delta \to 0$. The answer in general is no, according to the Bakushinskii's veto, which states that heuristic rules can not lead to convergence in the sense of worst case scenario for any regularization method [2]. Therefore, to guarantee a convergence result, one must impose certain conditions on the noisy data. We will use the following *noise condition.*

Assumption 3 $\{y^\delta\}$ is a family of noisy data satisfying $0 < \|y^\delta - y^\dagger\|_{\mathcal{Y}} \to 0$ as $\delta \to 0$ and there is a constant $\kappa > 0$ such that

$$\|y^\delta - F(x)\|_{\mathcal{Y}} \geq \kappa \|y^\delta - y^\dagger\|_{\mathcal{Y}}$$

for every y^δ and every $x \in S(y^\delta)$, where $S(y^\delta) := \{x_n^\delta : 0 \leq n \leq n_\infty\}$ generated by (4)–(5). □

Under Assumption 3, it is immediate to see that Rule 1 defines a finite integer n_*.

The noise condition was introduced in the seminal paper [7] for linear problems in Hilbert spaces. The reader may refer to [11, 18, 20] for recent developments using this condition, while a very comprehensive numerical study of heuristic rules, aside from the Hanke-Raus rule, for nonlinear Landweber iteration can be found in [8].

3 Analysis of Rule 1

We discuss the well-definedness of rule 1, i.e.,

$$\Theta(n_*(y^\delta), y^\delta) \to 0 \quad \text{as } \delta \to 0 \tag{9}$$

To achieve this result, we need an auxiliary integer $\hat{n}_\delta$ defined by slightly modifying the discrepancy principle, and establish that $\hat{n}_\delta \to \infty$ as $\delta \to 0$. This is an important step in establishing (9), which is stated formally below.

Lemma 19.1 *Under standard assumptions, let $\mu_0 > 0$ be chosen such that*

$$c_1 := 1 - \eta - \frac{1}{p^*}\left(\frac{\mu_0}{2c_0}\right)^{p^*-1} > 0. \tag{10}$$

Let $\{y^\delta\}$ be a family of noisy data satisfying Assumption 3. Let $n_* := n_*(y^\delta)$ be the integer defined by rule 1. Then $\Theta(n_*(y^\delta), y^\delta) \to 0$ as $\delta \to 0$. Consequently, $\left\|F(x^\delta_{n_*(y^\delta)}) - y^\delta\right\|_{\mathcal{Y}} \to 0$ and $n_*(y^\delta)\|y^\delta - y^\dagger\|_{\mathcal{Y}}^p \to 0$ as $\delta \to 0$.

Proof This is [18, Lemma 2.2]. □

Furthermore, Lemma 19.1 is key to establishing the well-definedness of the itertion (4)–(5) under rule 1.

Lemma 19.2 *Under the same assumptions in Lemma 19.1, let* $\{y^\delta\}$ *be a family of noisy data satisfying Assumption 3. Let* $n_* := n_*(y^\delta)$ *be the integer defined by Rule 1. Then* $x_n^\delta \in B_{3\rho}(x_0)$ *for all* $0 \le n \le n_*$ *if* δ *is sufficiently small.*

Aside from the previous lemma, we also need the convergence of the Landweber iteration (4)–(5) under noise-free environment. We refer the reader to [17, Lemma 2.5].

4 Main Result

Now we are about to prove the main result of this paper. We need two stability results based also on Assumptions 1 and 2.

Lemma 19.3 *Under same assumptions in Lemma 19.1, let* $\{y^\delta\}$ *be a family of noisy data satisfying Assumption 3. Let* $n_* := n_*(y^\delta)$ *be the integer defined by Rule 1.*

1. *If Assumption 1 holds and* $1 < r < \infty$, *then there is a sequence* $\{(\xi_n, x_n)\} \in \mathcal{G}01(\xi_0, x_0)$ *such that*

$$x_n^\delta \to x_n \quad \textit{and} \quad \xi_n^\delta \to \xi_n \quad \textit{as } \delta \to 0$$

for all $0 \le n \le \liminf_{\delta \to 0} n_*(y^\delta)$.

2. *If Assumption 2 holds and* $1 \le r < \infty$, *then for any subsequence* $\{y^{\delta_l}\}$, *with* $\delta_l \to 0$ *as* $l \to \infty$, *of* $\{y^\delta\}$, *by taking a subsequence if necessary, there is a sequence* $\{(\xi_n, x_n)\} \in \Gamma_{\mu_0,\mu_1}(\xi_0, x_0)$ *such that*

$$x_n^{\delta_l} \to x_n \quad \textit{and} \quad \xi_n^{\delta_l} \to \xi_n \quad \textit{as } l \to \infty$$

for all $0 \le n \le \liminf_{l \to \infty} n_*(y^{\delta_l})$.

Proof See [18, Lemma 2.5], [13, Lemma 3.8], as well as [17, Lemma 2.3]. □

And we have the main result.

Theorem 19.1 *Under same assumptions in Lemma 19.1, let* $\{y^\delta\}$ *be a sequence of noisy data satisfying Assumption 3 and let* $n_*(y^\delta)$ *be the integer determined by Rule 1.*

1. *If Assumption 1 holds and* $1 < r < \infty$, *then there is a solution* x_* *of* (1) *such that*

$$\left\| x^\delta_{n_*(y^\delta)} - x_* \right\|_{\mathcal{X}} \to 0 \quad \textit{and} \quad D_{\xi^\delta_{n_*(y^\delta)}} \mathcal{R}(x_*, x^\delta_{n_*(y^\delta)}) \to 0$$

as $\delta \to 0$. *If in addition* $\mathcal{N}(L(x^\dagger)) \subset \mathcal{N}(L(x))$ *for all* $x \in B_{3\rho}(x_0)$, *then* $x_* = x^\dagger$.

2. *If Assumption 2 holds and* $1 \le r < \infty$*, then for any subsequence* $\{y^{\delta_l}\}$*, with* $\delta_l \to 0$ *as* $l \to \infty$*, of* $\{y^{\delta}\}$*, by taking a subsequence of* $\{y^{\delta_l}\}$ *if necessary, there hold*

$$\left\| x^{\delta_l}_{n_*(y^{\delta_l})} - x_* \right\|_X \to 0 \quad \textit{and} \quad D_{\xi^{\delta_l}_{n_*(y^{\delta_l})}} \mathcal{R}(x_*, x^{\delta_l}_{n_*(y^{\delta_l})}) \to 0$$

as $l \to \infty$ *for some solution* x_* *of* (1) *in* $\mathcal{B}_{3\rho}(x_0)$*. If in addition* $\mathcal{N}(L(x^\dagger)) \subset \mathcal{N}(L(x))$ *for all* $x \in B_{3\rho}(x_0)$*, then*

$$\left\| x^{\delta}_{n_*(y^{\delta})} - x^\dagger \right\|_X \to 0 \quad \textit{and} \quad D_{\xi^{\delta}_{n_*(y^{\delta})}} \mathcal{R}(x^\dagger, x^{\delta}_{n_*(y^{\delta})}) \to 0$$

as $\delta \to 0$.

Proof See [18, Theorem 2.6]. □

Acknowledgements The author would like to thank the Ghent Analysis and PDE Center for the warm welcome, especially to Karel van Bockstal and Michael Ruzhansky.

References

1. Aster, R.C., Borchers, B., Thurber, C.H.: Parameter estimation and inverse problems, vol. 90. Elsevier Academic Press, Amsterdam (2005)
2. Bakushinskii, A.B.: Remarks on choosing a regularization parameter using the quasioptimality and ratio criterion. USSR Comput. Math. Math. Phys. **24**(4), 181–182 (1984)
3. BoÅ£, R.I., Hein, T.: Iterative regularization with a general penalty term–theory and application to L^1 and TV regularization. Inverse Probl. **28**(10), 104010 (2012)
4. Clason, C.: L^∞ fitting for inverse problems with uniform noise. Inverse Probl. **28**(10), 104007 (2012)
5. Clason, C., Nhu, V.H.: Bouligand-Landweber iteration for a non-smooth ill-posed problem. Numer. Math. **142**(4), 789–832 (2019)
6. Fu, Z., Jin, Q., Zhang, Z., Han, B., Chen, Y.: Analysis of a heuristic rule for the IRGNM in Banach spaces with convex regularization terms. Inverse Probl. **36**(7), 75002 (2020)
7. Hanke, M., Raus, T.: A general heuristic for choosing the regularization parameter in ill-posed problems. SIAM J. Sci. Comput. **17**(4), 956–972 (1996)
8. Hubmer, S., Sherina, E., Kindermann, S., Raik, K.: A numerical comparison of some heuristic stopping rules for nonlinear Landweber iteration. Electron. Trans. Numer. Anal. **57**, 216–241 (2022)
9. Jin, B., Lorenz, D.A.: Heuristic parameter-choice rules for convex variational regularization based on error estimates. SIAM J. Numer. Anal. **48**(3), 1208–1229 (2010)
10. Jin, Q.: Landweber-Kaczmarz method in Banach spaces with inexact inner solvers. Inverse Probl. **32**(10), 104005 (2016)
11. Jin, Q.: Hanke-Raus heuristic rule for variational regularization in Banach spaces. Inverse Probl. **32** (8), (2016)
12. Jin, Q.: On a heuristic stopping rule for the regularization of inverse problems by the augmented Lagrangian method. Numer. Math. **136**(4), 973–992 (2017)
13. Jin, Q., Wang, W.: Landweber iteration of Kaczmarz type with general non-smooth convex penalty functionals. Inverse Probl. **29**(8), 085011 (2013)

14. Jin, Q., Wang, W.: Analysis of the iteratively regularized Gauss-Newton method under a heuristic rule. Inverse Probl. **34**(3), (2018)
15. Kaltenbacher, B., Schöpfer, F., Schuster, T.: Iterative methods for nonlinear inverse ill-posed problems in Banach spaces: convergence and applications to parameter identification problems. Inverse Probl. **25**(6), (2009)
16. Liu, H., Real, R., Lu, X., Jia, X., Jin, Q.: Heuristic discrepancy principle for variational regularization of inverse problems. Inverse Probl. **36**(7), (2020)
17. Real, R., Jin, Q.: A revisit on Landweber iteration. Inverse Probl. **36**(7), 75011 (2020)
18. Real, R.: Hanke-Raus rule for Landweber iteration in Banach spaces. Numer. Math. **156**(1), 345–373 (2024)
19. Schöpfer, F., Louis, A.K., Schuster, T.: Nonlinear iterative methods for linear ill-posed problems in Banach spaces. Inverse Probl. **22**(1), (2006)
20. Zhang, Z., Jin, Q.: Heuristic rule for non-stationary iterated Tikhonov regularization in Banach spaces. Inverse Probl. **34**(11), 115002 (2018)

On the q-Analogue of Duhamel's Principle

Irfan Ali, Mohammed Elamine Sebih, and Serikbol Shaimardan

Abstract In this short note, we discuss the q-analogue of Duhamel's principle for q-evolution equations of order $k \geq 1$, generated by Jackson's operator.

Keywords Evolution equations · Non-homogeneous problems · Duhamel's principle · q-calculus

MSC2020 34A99 · 05A30 · 33D05

1 Introduction

Duhamel's principle is a key in proving the existence and uniqueness of solutions to non-homogeneous partial differential equations. Numerous equations, including the wave equation, the heat equation and the Scrödinger equation are studied using this principle, see e.g. [2, 4] for more details and applications. However, a carefully written proof of it seems, surprisingly, to be missing in the literature. Our aim in this short note is to fill this gap, and to provide the proof for the q-analogue of this principle for specific versions of q-evolution equations generated by Jackson's operator. The paper is organised as follows: In the next section, we give the necessary preliminaries about quantum calculus. For the convenience of the reader, Duhamel's principle in the classical case is included in Sect. 3. In Sect. 4, we present our results.

I. Ali · S. Shaimardan
Department of Mathematics: Analysis, Logic and Discrete Mathematics, Ghent University, Ghent, Belgium

M. E. Sebih (✉)
Laboratory of Geomatics, Ecology and Environment (LGEO2E),
Mustapha Stambouli University of Mascara, Mascara, Algeria
e-mail: ma.sebih@univ-mascara.dz

M. Chatzakou et al. (eds.), *Direct and Inverse Problems with Applications*,
Trends in Mathematics 12, https://doi.org/10.1007/978-3-031-98645-1_20

2 Preliminaries

In this section, we recall some notations related to the q-calculus. We will always assume that $0 < q < 1$. The q-real number $[\alpha]_q$ is defined by

$$[\alpha]_q := \frac{1-q^\alpha}{1-q}.$$

We define the q-lattice $\mathbb{R}_q \subset \mathbb{R}$ by

$$\mathbb{R}_q := \{x^k : x = \pm q^k;\ k \in \mathbb{Z}\},$$

and accordingly, for $a, b \in \mathbb{R}$,

$$[a, b]_q := [a, b] \cap \mathbb{R}_q.$$

The q-analogue differential operator $D_q f(x)$ (also called Jackson's operator) is defined by

$$D_q f(x) = \frac{f(x) - f(qx)}{x(1-q)}. \tag{1}$$

The q-derivative of a product of two functions has the form

$$D_q(fg)(x) = f(qx)D_q(g)(x) + D_q(f)(x)g(x),$$

and the q^2-derivative (called Rubin's operator) is defined by (see [6] and [7])

$$\partial_q f(x) = \frac{f(q^{-1}x) + f(-q^{-1}x) - f(qx) + f(-qx) - 2f(-x)}{2x(1-q)}. \tag{2}$$

Jackson's and Rubin's operators are related by

$$\partial_q f(x) = D_q f(x) + D_{q^{-1}} f(x).$$

The q-integral (or Jackson integral) is defined by (see [5])

$$\int_0^x f(t)d_qt = (1-q)x \sum_{m=0}^{\infty} q^m f(xq^m),$$

and, more generally,

$$\int_a^b f(x)d_qx = \int_0^b f(x)d_qx - \int_0^a f(x)d_qx,$$

provided the sums converge absolutely. Note that

$$\int_a^x D_q f(t) d_q t = f(x) - f(a).$$

We refer the reader to [1, 3] for more details about the q-calculus.

In the following lemma, we give the q-version of the differentiation of functions defined in terms of integrals depending on a parameter.

Lemma 1 *Let $f : \mathbb{R}^2 \to \mathbb{R}$ be a function such that $D_{q,x} f(x, y)$ exists. Then, the function F defined on $\mathbb{R}$ by $F(x) = \int_0^x f(x, t) d_q t$ is q-differentiable and we have*

$$D_q F(x) = f(qx, x) + \int_0^x D_{q,x} f(x, t) d_q t,$$

or equivalently

$$D_q F(x) = f(x, x) + \int_0^{qx} D_{q,x} f(x, t) d_q t.$$

3 Classical Duhamel's Principle

The following variants of Duhamel's principle are commonly used.

3.1 First Order Duhamel's Principle

Let us consider the following Cauchy problem for the first-order non-homogeneous linear evolution equation

$$\begin{cases} u_t(t, x) - Lu(t, x) = f(t, x), & (t, x) \in (0, \infty) \times \mathbb{R}^d, \\ u(0, x) = u_0(x), & x \in \mathbb{R}^d, \end{cases}$$

where L is a linear differential operator that does not include time derivatives.

Theorem 1 *The solution to the Cauchy problem (3.1) is given by*

$$u(t, x) = w(t, x) + \int_0^t v(t, x; s) ds,$$

where $w(t,x)$ is the solution to the homogeneous problem

$$\begin{cases} w_t(t,x) - Lw(t,x) = 0, & (t,x) \in (0,\infty) \times \mathbb{R}^d, \\ w(0,x) = u_0(x), & x \in \mathbb{R}^d, \end{cases}$$

and $v(t,x;s)$ solves the auxiliary Cauchy problem

$$\begin{cases} v_t(t,x;s) - Lv(t,x;s) = 0, & (t,x) \in (s,\infty) \times \mathbb{R}^d, \\ v(s,x;s) = f(s,x), & x \in \mathbb{R}^d, \end{cases}$$

where s is a time-like parameter varying over $(0,\infty)$.

3.2 Second Order Duhamel's Principle

Let us consider the Cauchy problem for the second-order non-homogeneous linear evolution equation

$$\begin{cases} u_{tt}(t,x) - Lu(t,x) = f(t,x), & (t,x) \in (0,\infty) \times \mathbb{R}^d, \\ u(0,x) = u_0(x), \quad u_t(0,x) = u_1(x), & x \in \mathbb{R}^d, \end{cases}$$

for some linear partial differential operator L over the space variable x.

Theorem 2 *The solution to the Cauchy problem (3.2) is given by*

$$u(t,x) = w(t,x) + \int_0^t v(t,x;s)ds,$$

where $w(t,x)$ is the solution to the homogeneous problem

$$\begin{cases} w_{tt}(t,x) - Lw(t,x) = 0, & (t,x) \in (0,\infty) \times \mathbb{R}^d, \\ w(0,x) = u_0(x), \quad w_t(0,x) = u_1(x), & x \in \mathbb{R}^d, \end{cases}$$

and $v(t,x;s)$ solves the auxiliary Cauchy problem

$$\begin{cases} v_{tt}(t,x;s) - Lv(t,x;s) = 0, & (t,x) \in (s,\infty) \times \mathbb{R}^d, \\ v(s,x;s) = 0, \quad v_t(s,x;s) = f(s,x), & x \in \mathbb{R}^d, \end{cases}$$

where s is a time-like parameter varying over $(0,\infty)$.

3.3 k-order Duhamel's Principle

More generally, if we consider the Cauchy problem for the k-order non-homogeneous linear evolution equation

$$\begin{cases} \partial_t^k u(t,x) - Lu(t,x) = f(t,x), & (t,x) \in (0,\infty) \times \mathbb{R}^d, \\ \partial_t^j u(0,x) = u_j(x), \text{ for } \; j = 0, \cdots, k-1, & x \in \mathbb{R}^d, \end{cases} \tag{3}$$

then, Duhamel's principle reads as follows:

Theorem 3 *The solution to the Cauchy problem* (3) *has the representation*

$$u(t,x) = w(t,x) + \int_0^t v(t,x;s)ds,$$

where $w(t,x)$ *is the solution to the homogeneous problem*

$$\begin{cases} \partial_t^k w(t,x) - Lw(t,x) = 0, & (t,x) \in (0,\infty) \times \mathbb{R}^d, \\ \partial_t^j w(0,x) = u_j(x), \textit{ for } \; j = 0, \cdots, k-1, & x \in \mathbb{R}^d, \end{cases}$$

and $v(t,x;s)$ *solves*

$$\begin{cases} \partial_t^k v(t,x;s) - Lv(t,x;s) = 0, & (t,x) \in (s,\infty) \times \mathbb{R}^d, \\ \partial_t^j v(s,x;s) = 0, \textit{ for } \; j = 0, \cdots, k-2, \; \partial_t^{k-1} v(s,x;s) = f(s,x), & x \in \mathbb{R}^d, \end{cases}$$

where $s \in (0,\infty)$.

4 q-Analogue of Duhamel's Principle

In this section, we present the q-analogue of the last versions of Duhamel's principle.

4.1 First Order Duhamel's Principle for q-Evolution Equations

Let us consider the following Cauchy problem for the first-order non-homogeneous linear q-evolution equation generated by Jackson's operator:

$$\begin{cases} D_{q,t}u(t,x) - L_{q,x}u(t,x) = f(t,x), & (t,x) \in (0,\infty)_q \times \mathbb{R}_q, \\ u(0,x) = u_0(x), \quad x \in \mathbb{R}_q, \end{cases}$$

where $L_{q,x}$ is a linear q-differential operator that involves no time derivatives. Then, Duhamel's principle reads:

Theorem 4 *The solution to the Cauchy problem (4.1) is given by*

$$u(t,x) = w(t,x) + \int_0^t v(t,x;s)d_q s,$$

where $w(t,x)$ is the solution to the homogeneous problem

$$\begin{cases} D_{q,t}w(t,x) - L_{q,x}w(t,x) = 0, & (t,x) \in (0,\infty)_q \times \mathbb{R}_q, \\ w(0,x) = u_0(x), \quad x \in \mathbb{R}_q, \end{cases}$$

and $v(t,x;s)$ solves the auxiliary Cauchy problem

$$\begin{cases} D_{q,t}v(t,x;s) - L_{q,x}v(t,x;s) = 0, & (t,x) \in (s,\infty)_q \times \mathbb{R}_q, \\ v(qs,x;s) = f(s,x), \quad x \in \mathbb{R}_q, \end{cases}$$

where s is a time-like parameter varying over $(0,\infty)_q$.

4.2 Second Order Duhamel's Principle for q-Evolution Equations

Let us consider the Cauchy problem for the second-order non-homogeneous linear q-evolution equation

$$\begin{cases} D_{q,tt}u(t,x) - L_{q,x}u(t,x) = f(t,x), & (t,x) \in (0,\infty)_q \times \mathbb{R}_q, \\ u(0,x) = u_0(x), \quad D_{q,t}u(0,x) = u_1(x), \quad x \in \mathbb{R}_q, \end{cases}$$

for some linear q-differential operator $L_{q,x}$ over the space variable x.

Theorem 5 *The solution to the Cauchy problem (4.2) is given by*

$$u(t,x) = w(t,x) + \int_0^t v(t,x;s)d_q s,$$

where $w(t,x)$ is the solution to the homogeneous problem

$$\begin{cases} D_{q,tt}w(t,x) - L_{q,x}w(t,x) = 0, & (t,x) \in (0,\infty)_q \times \mathbb{R}_q, \\ w(0,x) = u_0(x), \quad D_{q,t}w(0,x) = u_1(x), \quad x \in \mathbb{R}_q, \end{cases}$$

and $v(t, x; s)$ solves the auxiliary Cauchy problem

$$\begin{cases} D_{q,tt}v(t, x; s) - L_{q,x}v(t, x; s) = 0, & (t, x) \in (s, \infty)_q \times \mathbb{R}_q, \\ v(qs, x; s) = 0, \quad D_{q,t}v(qs, x; s) = f(s, x), & x \in \mathbb{R}_q, \end{cases}$$

where s is a time-like parameter varying over $(0, \infty)_q$.

4.3 k-Order Duhamel's Principle for q-Evolution Equations

If we consider the following Cauchy problem for the k-order non-homogeneous q-evolution equation

$$\begin{cases} D_{q,t}^{(k)}u(t, x) - L_{q,x}u(t, x) = f(t, x), & (t, x) \in (0, \infty)_q \times \mathbb{R}_q, \\ D_{q,t}^{(j)}u(0, x) = u_j(x), \text{ for } \; j = 0, \cdots, k-1, & x \in \mathbb{R}_q, \end{cases} \tag{4}$$

then, the principle reads as follows:

Theorem 6 *The solution to the Cauchy problem* (4) *is given by*

$$u(t, x) = w(t, x) + \int_0^t v(t, x; s) d_q s,$$

where $w(t, x)$ is the solution to the homogeneous problem

$$\begin{cases} D_{q,t}^{(k)}w(t, x) - L_{q,x}w(t, x) = 0, & (t, x) \in (0, \infty)_q \times \mathbb{R}_q, \\ D_{q,t}^{(j)}w(0, x) = w_j(x), \textit{ for } \; j = 0, \cdots, k-1, & x \in \mathbb{R}_q, \end{cases}$$

and $v(t, x; s)$ solves the auxiliary Cauchy problem

$$\begin{cases} D_{q,t}^{(k)}v(t, x; s) - L_{q,x}v(t, x; s) = 0, \quad (t, x) \in (s, \infty)_q \times \mathbb{R}_q, \\ D_{q,t}^{j}v(qs, x; s) = 0, \textit{ for } \; j = 0, \cdots, k-2, \; D_{q,t}^{(k-1)}v(qs, x; s) = f(s, x), \quad x \in \mathbb{R}_q, \end{cases}$$

where s is a fixed parameter varying over $(0, \infty)_q$.

Remark 1 In Theorems 4, 5 and 6, the q-differential operator $L_{q,x}$ can be either generated by Jackson's operator (1) or Rubin's operator (2). The results remain true.

Remark 2 We easily see that if $q \to 1^-$, then we recapture the classical Duhamel's principles.

Acknowledgements This research was funded by the FWO Odysseus 1 grant G.0H94.18N: Analysis and Partial Differential Equations, and by the Methusalem programme of the Ghent University Special Research Fund (BOF) (Grant number 01M01021).

References

1. Cheung, P., Kac, V.: Quantum calculus. Edwards Brothers. Inc., Ann Arbor, MI, USA (2020)
2. Ebert, M.R., Reissig, M.: Methods for partial differential equations. Birkhäuser (2018)
3. Ernst, T.: A comprehensive treatment of q-calculus. Birkhäuser/Springer Basel AG, Basel (2012)
4. Evans, L.C.: Partial differential equations. American Mathematical Society (1998)
5. Jackson, F.H.: On a q-definite integrals. Quart. J. Pure and Appl. Math. **41**, 193–203 (1910)
6. Rubin, R.L.: A q^2-analogue operator for q^2-analogue Fourier analysis. J. Math. Anal. Appl. **212**, 571–582 (1997)
7. Rubin, R.L.: Duhamel solutions of non-homogeneous q^2-analogue wave equations. Proc. Amer. Math. Soc. **135**, 777–785 (2007)

Nonlinear Diffusion, and Geometric and Functional Inequalities on Smooth Metric Measure Spaces

Ali Taheri

Abstract This extended abstract is based on a talk given at the workshop and summer school "Direct and Inverse Problems with Applications" in the Ghent Analysis and PDE Center in August 2024. I take a look at nonlinear diffusion equations of slow and fast types and their links with some geometric and functional inequalities in the framework of smooth metric measure spaces. The article presents some introduction and background in a summer school style as well as new results.

Keywords Functional inequalities · Isoperimetric inequalities · Smooth metric measure spaces · Nonlinear diffusion · Entropy

MSC2020 53C44 · 58J60 · 58J35 · 60J60

1 Introduction and Preliminaries

Inequalities between some notion of "*size*" of sets and their perimeters, functions and their gradients, or more generally, tensor fields and their differentials, covariant or Lie derivatives, play an important role in geometric analysis. I will give an introduction to certain aspects of the subject (without proofs) and describe some new research directions emerging from the need to extend known inequalities from Euclidean to the framework of (smooth) metric measure spaces (*see*, e.g., [1–3, 7, 9–11, 18–20]).

A. Taheri (✉)
School of Mathematical and Physical Sciences, University of Sussex, Falmer, United Kingdom
e-mail: A.Taheri@sussex.ac.uk

M. Chatzakou et al. (eds.), *Direct and Inverse Problems with Applications*, Trends in Mathematics 12, https://doi.org/10.1007/978-3-031-98645-1_21

2 Isoperimetric Inequality and the Isoperimetric Profile

Recall that the classical isoperimetric inequality in $\mathbb{R}^n$ $(n \geq 2)$ asserts that for any sufficiently smooth compact subdomain $\Omega \subset \mathbb{R}^n$:

$$n\omega_n^{1/n}[\mathrm{Vol}(\Omega)]^{1-1/n} \leq \mathrm{Per}(\Omega).$$

Here $\mathrm{Vol}(\Omega) = \mathscr{L}^n(\Omega)$ denotes the volume of Ω, $\mathrm{Per}(\Omega) = \mathscr{H}^{n-1}(\partial\Omega)$ denotes the perimeter of Ω and ω_n is the volume of the unit n-ball. This inequality can equivalently be written (with $\mathbb{B} = \mathbb{B}_n$ denoting the unit n-ball) in the more suggestive form

$$\frac{\mathrm{Per}(\Omega)}{[\mathrm{Vol}(\Omega)]^{\frac{n-1}{n}}} \geq \frac{\mathrm{Per}(\mathbb{B})}{[\mathrm{Vol}(\mathbb{B})]^{\frac{n-1}{n}}} = \frac{n\omega_n}{\omega_n^{\frac{n-1}{n}}} = n\omega_n^{\frac{1}{n}}. \tag{1}$$

The inequality asserts that for fixed volume, balls *minimise* perimeter, or for fixed perimeter, balls *maximise* volume. There are various ways to prove this inequality, but the most common and quickest one is via the Brunn-Minkowski inequality:

$$[\mathrm{Vol}(A+B)]^{\frac{1}{n}} \geq [\mathrm{Vol}(A)]^{\frac{1}{n}} + [\mathrm{Vol}(B)]^{\frac{1}{n}},$$

where $A, B \subset \mathbb{R}^n$ are non-empty compact sets and $A + B = \{a + b \in \mathbb{R}^n : a \in A, b \in B\}$. Another way of writing (1) is to introduce for $v \geq 0$ the isoperimetric profile function $I(v) = \inf\{\mathrm{Per}(\Omega) : \mathrm{Vol}(\Omega) = v\}$. Then $\mathrm{Per}(\Omega) \geq I(\mathrm{Vol}(\Omega))$ where $I(v) = n\omega_n^{1/n} v^{1-1/n}$. Moreover, for any n-ball $\mathcal{B}$ we have $I(\mathrm{Vol}(\mathcal{B})) = \mathrm{Per}(\mathcal{B})$.

The Levy-Gromov isoperimetric inequality. A first natural question here is if the isoperimetric inequality has a counterpart for Riemannian manifolds, and if so, what is it? Starting with the cases of the round unit n-sphere $\mathbb{S}^n$, and the real hyperbolic n-space $\mathbb{H}^n$ (i.e., model spaces of constant sectional curvatures ± 1) we have the following, which are in spirit quite close to (and complement) what was observed above for $\mathbb{R}^n$:

- Given $\Omega \subset \mathbb{S}^n$, if $\mathcal{B} \subset \mathbb{S}^n$ is a geodesic ball (i.e., a spherical cap) such that $\mathrm{Vol}(\Omega) = \mathrm{Vol}(\mathcal{B})$ then $\mathrm{Per}(\Omega) \geq \mathrm{Per}(\mathcal{B})$.
- Given $\Omega \subset \mathbb{H}^n$ with compact closure, if $\mathcal{B} \subset \mathbb{H}^n$ is a geodesic ball such that $\mathrm{Vol}(\Omega) = \mathrm{Vol}(\mathcal{B})$ then $\mathrm{Per}(\Omega) \geq \mathrm{Per}(\mathcal{B})$.

Returning now to the original question, the answer for compact manifolds without boundary (i.e., closed manifolds) is in the affirmative and is provided by the celebrated Levy-Gromov isoperimetric inequality:

- Let $(\mathscr{M}, g)$ be a closed Riemannian manifold with $\dim(\mathscr{M}) = n \geq 2, \mathscr{R}ic(\mathscr{M}) \geq n-1[= \mathscr{R}ic(\mathbb{S}^n)]$. Let $\Omega \subset \mathscr{M}$ be a bounded smooth subdomain and let $\mathcal{B} \subset \mathbb{S}^n$ be a round geodesic open ball (spherical cap). Then we have the implication

$$\frac{\mathrm{Vol}(\Omega)}{\mathrm{Vol}(\mathscr{M})} = \frac{\mathrm{Vol}(\mathscr{B})}{\mathrm{Vol}(\mathbb{S}^n)} \implies \frac{\mathrm{Per}(\Omega)}{\mathrm{Vol}(\mathscr{M})} \geq \frac{\mathrm{Per}(\mathscr{B})}{\mathrm{Vol}(\mathbb{S}^n)}.$$

Here $\mathscr{M}$ can be *any* closed manifold and the result compares the isoperimetric nature of $\mathscr{M}$ to that of an associated model space (of constant curvature).

This discussioin can be conveniently reformulated in terms of an object called the isoperimetric profile (of the Riemannian manifold). To this end, for $0 \leq v \leq 1$ let

$$I_{(\mathscr{M},g)}(v) = \inf \left\{ \frac{\mathrm{Per}(\Omega)}{\mathrm{Vol}(\mathscr{M})} : \frac{\mathrm{Vol}(\Omega)}{\mathrm{Vol}(\mathscr{M})} = v \right\}.$$

Then the Levy-Gromov isoperimetric inequality is the statement that $I_{(\mathscr{M},g)} \geq I_{(\mathbb{S}^n, g_{\mathbb{S}^n})}$.

3 The Gauss Space $(\mathbb{R}^n, g_{\mathbb{R}^n}, d\gamma_n)$

The Gauss space is the most fundamental and natural example of a smooth metric measure space (that we will come to shortly). It consists of the Euclidean space $\mathbb{R}^n$ ($n \geq 1$) equipped with its canonical metric $g_{\mathbb{R}^n}$ and the Gaussian measure $d\gamma = d\gamma_n$

$$d\gamma_n = \prod_{j=1}^{n} \frac{e^{-\frac{1}{2}x_j^2}}{\sqrt{2\pi}} dx_j = \frac{e^{-\frac{1}{2}|x|^2}}{(2\pi)^{\frac{n}{2}}} dx. \tag{2}$$

In the study of isoperimetric inequality for Gauss space, *half-spaces* play an important role. For unit vector $u \in \mathbb{R}^n$ and $a \in \mathbb{R}$, let $\mathscr{H}_a = \{x \in \mathbb{R}^n : \langle x, u \rangle \leq a\}$. Moreover, for any Borel measurable set $A \subset \mathbb{R}^n$ set

$$\gamma_n^+(A) = \liminf_{\varepsilon \searrow 0} \frac{\gamma_n(A_\varepsilon) - \gamma_n(A)}{\varepsilon}, \qquad \gamma_n(A) = \int_A d\gamma_n,$$

where for $\varepsilon > 0$ we write $A_\varepsilon = \{x \in \mathbb{R}^n | \mathrm{dist}(x, A) < \varepsilon\}$. Here γ_n^+ is called Minkowski's exterior boundary measure (or content or simply perimeter). Then we have

$$\gamma_n(A) = \gamma_n(\mathscr{H}_a) \implies \gamma_n^+(A) \geq \gamma_n^+(\mathscr{H}_a).$$

Thus among all sets of a fixed Gaussian measure, half-spaces have the least perimeter.

As with the Levy-Gromov case, one can introduce the isoperimetric profile by writing, $I(v) = \inf\{\gamma_n^+(A) : \gamma_n(A) = v\}$ for $v \in [0, 1]$. The Gaussian isoperimetric inequality then asserts that

$$I(\gamma_n(A)) \leq \gamma_n^+(A). \tag{3}$$

Here one can also show that (regardless of n) $I = \phi \circ \Phi^{-1} : [0, 1] \to [0, 1/\sqrt{2\pi}]$ where

$$\Phi(r) = \gamma_1(-\infty, r) = \int_{-\infty}^{r} \frac{e^{-\frac{1}{2}x^2}}{\sqrt{2\pi}} dx, \qquad \phi = \Phi'.$$

Note that for half-spaces we have $\Phi(a) = \gamma_n(\mathscr{H}_a)$ and $\phi(a) = \gamma_n^+(\mathscr{H}_a)$.

Functional inequalities on $(\mathbb{R}^n, g_{\mathbb{R}^n}, d\gamma_n)$. With this introduction on isoperimetric inequalities, and still within the context of the Gauss space $(\mathbb{R}^n, g_{\mathbb{R}^n}, d\gamma_n)$, we present two important functional inequalities, that will serve as motivation for the further development of this article. These are in turn the Poincaré inequality (also called the spectral gap inequality for the Hermite polynomials) and the logarithmic Sobolev inequality. (See, e.g., [3, 18] and the references therein for more.)

- The first inequality takes the form:

$$\int_{\mathbb{R}^n} f^2 \, d\gamma_n - \left(\int_{\mathbb{R}^n} f \, d\gamma_n \right)^2 \leq \int_{\mathbb{R}^n} |\nabla f|^2 d\gamma_n. \tag{4}$$

- The second inequality, i.e., the logarithmic Sobolev inequality takes the form:

$$\int_{\mathbb{R}^n} |f|^2 \log |f|^2 \, d\gamma_n - \int_{\mathbb{R}^n} |f|^2 \, d\gamma_n \log \int_{\mathbb{R}^n} |f|^2 \, d\gamma_n \leq 2 \int_{\mathbb{R}^n} |\nabla f|^2 \, d\gamma_n, \tag{5}$$

or equivalently

$$\int_{\mathbb{R}^n} |f|^2 \log \frac{|f|^2}{||f||^2_{L^2(d\gamma_n)}} \, d\gamma_n \leq 2 \int_{\mathbb{R}^n} |\nabla f|^2 \, d\gamma_n.$$

Logarithmic Sobolev inequalities are well-known to be of huge significance and to have numerous applications. In the context described above, one important implication is the following hypercontractivity of the Nelson semigroup $P_t = e^{tN}$ $(1 < p < 2)$:[1]

$$||e^{tN} f||^2_{L^2(d\gamma_n)} = \int_{\mathbb{R}^n} |e^{tN} f|^2 \, d\gamma_n \leq ||f||^2_{L^p(d\gamma_n)}, \qquad e^{-t} \leq \sqrt{p-1}.$$

In relation to the discussion on logarithmic Sobolev inequalities above there two useful points to remember.

[1] One interesting application of this hypercontractivity is to polynomial chaos.

- Firstly, the function $r \mapsto \phi(r) = \log ||f||_{L^{1/r}(d\mu)}$ for $0 < r < 1$ is convex.
- Secondly, the derivative of the L^p-norm with respect to p is given by the formula:

$$\begin{aligned}\frac{d}{dp}\|f\|_{L^p(d\mu)} &= \frac{1}{p^2}\|f\|_{L^p(d\mu)}^{1-p}\left[\int_{\mathscr{M}} |f|^p \log |f|^p \, d\mu - \int_{\mathscr{M}} |f|^p \, d\mu \log \int_{\mathscr{M}} |f|^p \, d\mu\right] \\ &= \frac{1}{p^2}\|f\|_{L^p(d\mu)}^{1-p} \int_{\mathscr{M}} |f|^p \log \frac{|f|^p}{||f||_{L^p(d\mu)}^p} \, d\mu. \end{aligned} \tag{6}$$

Notation. We make use of the following notations for Energy, Fisher Information, Variance and Entropy of a function f on $(\mathscr{M}, g, d\mu)$. Here we assume $f > 0$. Moreover, later we often take $d\mu$ a probability measure, although for the sake of these notations, this restriction is not necessary.

$$\mathrm{E}_\mu(f) = \int_{\mathscr{M}} |\nabla f|^2 \, d\mu, \qquad I_\mu(f) = \int_{\mathscr{M}} \frac{|\nabla f|^2}{f} \, d\mu,$$

$$\mathrm{Var}_\mu(f) = \int_{\mathscr{M}} f^2 \, d\mu - \left(\int_{\mathscr{M}} f \, d\mu\right)^2,$$

$$\mathrm{Ent}_\mu(f) = \left[\int_{\mathscr{M}} f \log f \, d\mu - \int_{\mathscr{M}} f \, d\mu \log \int_{\mathscr{M}} f \, d\mu\right].$$

The variance and entropy are easily seen to be non-negative, when $d\mu$ is a probability measure, by the use of Jensen's inequality. (For entropy use the convex function $\phi(t) = t \log t$ for $t > 0$.) With the notations above, we can re-write (4) and (5) as

$$\mathrm{Var}_{\gamma_n}(f) \leq \mathrm{E}_{\gamma_n}(f), \qquad \mathrm{Ent}_{\gamma_n}(f^2) \leq 2\mathrm{E}_{\gamma_n}(f),$$

respectively. Furthermore we can re-write (6) as

$$\frac{d}{dp}\|f\|_{L^p(d\mu)} = \frac{1}{p^2}\|f\|_{L^p(d\mu)}^{1-p}\mathrm{Ent}_\mu(|f|^p).$$

Geometric and Functional Inequalities on $\mathbb{R}^n$. It is natural and useful to list here (for purely motivational purposes) *some* of the geometric and functional inequalities on $\mathbb{R}^n$. Each of these inequalities underlines, signifies or links to a specific feature of $\mathbb{R}^n$. We also refrain from writing the word "*inequality*" each time for brevity. These are: Nash, Hardy, Poincaré, Faber-Krahn, Isoperimetric, Loomis-Whitney, Brascamp-Lieb, Hausdorff-Young, Logarithmic Sobolev, Caffarelli-Kohn-Nirenberg, Gagliardo-Nirenberg-Sobolev, Littlewood-Paley g-function, Hardy-Littlewood-Wiener maximal function, Isocapacitory, measure-capacity to name a few. (For more, *see* [1, 2, 4–6, 8–11, 18, 19] and the references therein.)

Stages of Generalisations. In passing from $\mathbb{R}^n$ to more general spaces, depending on the algebraic/geometric structure of the space involved or the technicalities available, one can present the following list:[2]

- Locally Compact Abeilan Groups
- Lie Groups, Homogenous Spaces, Symmetric Spaces
- Riemannian Manifolds $(\mathscr{M}, g)$
- Smooth Metric Measure Spaces $(\mathscr{M}, g, d\mu)$
- Metric Measure Spaces $(X, \mathsf{d}, \mathfrak{m})$ (No differentiable structure!)

4 Smooth Metric Measure Spaces $(\mathscr{M}, g, d\mu)$

A smooth metric measure space is a triple $(\mathscr{M}, g, d\mu)$ where $(\mathscr{M}, g)$ is a smooth Riemannian manifold of dimension $n \geq 2$, $d\mu = e^{-\varphi} dv_g$ is a positive weighted measure and dv_g is the Riemannian volume measure on $\mathscr{M}$. The diffusion operator associated with the triple $(\mathscr{M}, g, d\mu)$ is the φ-Laplacian (that depending on the context is also called the Witten, drifting or weighted Laplacian). It is defined for $w \in \mathscr{C}^2(\mathscr{M})$ by

$$\Delta_\varphi w = \Delta w - \langle \nabla \varphi, \nabla w \rangle. \tag{7}$$

As is readily seen it is a symmetric diffusion operator with respect to the invariant measure $d\mu$ and in the Riemannian context $\varphi \equiv 0$ (i.e., $d\mu = dv_g$) coincides with the usual Laplace-Beltrami operator.

The Ricci tensors $\mathscr{R}ic_\varphi^m(g)$ and the φ-Laplacian Δ_φ. The Bakry-Èmery m-Ricci curvature tensor $\mathscr{R}ic_\varphi^m(g)$ associated with the triple $(\mathscr{M}, g, d\mu)$ is defined as

$$\mathscr{R}ic_\varphi^m(g) = \mathscr{R}ic(g) + \nabla\nabla\varphi - \frac{\nabla\varphi \otimes \nabla\varphi}{m-n}, \qquad m \geq n. \tag{8}$$

Here $\mathscr{R}ic(g)$ is the usual Riemannain Ricci curvature tensor of g, $\nabla\nabla\varphi = \text{Hess}(\varphi)$ denotes the Hessian of φ, and $m \geq n$ is a constant (*see* [3]). For the sake of clarification, in (8), when $m = n$, by convention f is only allowed to be a constant, thus giving $\mathscr{R}ic_\varphi^n(g) = \mathscr{R}ic(g)$. By formally taking the limit $m \nearrow \infty$ in (8) we also define

$$\mathscr{R}ic_\varphi(g) = \mathscr{R}ic(g) + \nabla\nabla\varphi. \tag{9}$$

---

[2] The first two on the list are generalisations of $\mathbb{R}^n$ with a view towards maintaining the harmonic analysis tools. In the first case, one can essentially define the Fourier transform (Gelfand transform, characters and Pontryagin duality). In the second case, the situation is much more complicated, but there are more advanced tools of a similar nature available (Representation theory, Peter-Weyl theory, Harish-Chandra theory, Fourier-Helgason transform and Plancherel measure to name a few). As for the others, the set of techniques is vast. In the last case the theory of optimal transport is a key tool.

The significance of (9) lies in its appearance in the weighted Bochner-Weitzenböck formula, asserting that for any function w of class $\mathscr{C}^3(M)$ it holds,

$$\frac{1}{2}\Delta_\varphi|\nabla w|^2 - \langle\nabla w, \nabla\Delta_\varphi w\rangle = |\nabla\nabla w|^2 + \mathscr{R}ic_\varphi(\nabla w, \nabla w). \tag{10}$$

The above formula does not have a counterpart for $m < \infty$, however, by an application of Cauchy-Schwarz inequality, one can easily show that the following *inequality* holds for $\mathscr{R}ic^m_\varphi(g)$,

$$\frac{1}{2}\Delta_\varphi|\nabla w|^2 - \langle\nabla w, \nabla\Delta_\varphi w\rangle \geq \frac{1}{m}|\Delta_\varphi w|^2 + \mathscr{R}ic^m_\varphi(\nabla w, \nabla w). \tag{11}$$

Both these relations and the corresponding Ricci tensors will appear in the formulation of the curvature-dimension condition below. (See also [1, 3, 12–15] for more.)

Some examples of smooth metric measure spaces $(\mathscr{M}, g, d\mu)$ **with** $d\mu = e^{-\varphi}dv_g$. Let us proceed by giving some related and useful examples on smooth metric measure spaces to illustrate the scope of the theory.

- (Riemannian) $(\mathscr{M}, g)$ with constant φ: $\Delta_\varphi \equiv \Delta, \mathscr{R}ic^m_\varphi(g) = \mathscr{R}ic_\varphi(g) = \mathscr{R}ic(g)$.
- (Euclidean weighted space) $(\mathbb{R}^n, g_{\mathbb{R}^n}, d\mu = e^{-\varphi}dx)$: $\Delta_\varphi u = \Delta u - \langle\nabla\varphi, \nabla u\rangle$. Here $\mathscr{R}ic_\varphi(g) = \mathscr{R}ic(g) + \nabla\nabla\varphi = \nabla\nabla\varphi$ and so $\mathscr{R}ic_\varphi(g) \geq 0 \iff \varphi$ is convex. Furthermore from (8) we have

$$\mathscr{R}ic^m_\varphi(g) = \nabla\nabla\varphi - \frac{\nabla\varphi\otimes\nabla\varphi}{m-n}, \qquad n \leq m < \infty.$$

- (Ornstein-Uhlenbeck) $(\mathbb{R}^n, g_{\mathbb{R}^n}, d\gamma_n)$ with $d\gamma_n$ as in (2): $\Delta_\varphi u = \Delta u - \langle x, \nabla u\rangle$. Here $\mathscr{R}ic_\varphi(g) = \nabla\nabla\varphi = \mathrm{I}$. Moreover,

$$\mathscr{R}ic^m_\varphi(g) = \nabla\nabla\varphi - \frac{\nabla\varphi\otimes\nabla\varphi}{m-n} = \mathrm{I} - \frac{x\otimes x}{m-n}, \qquad n \leq m < \infty.$$

- (Interval case) $\mathscr{M} = (a, b)$, $d\mu = e^{-\varphi}dx$: $\Delta_\varphi u = u'' - \varphi' u'$. Here it is seen that

$$\mathscr{R}ic^m_\varphi(g) \geq k, \qquad when \qquad \varphi'' \geq k + \varphi'^2/(m-1).$$

  The cases $d\mu = (\sin t)^{m-1}\,dt$ on $(0, \pi)$ and $d\mu = (\sinh t)^{m-1}\,dt$ on $(0, \infty)$ arise in the study of isoperimetric inequality on $\mathbb{S}^n$ and $\mathbb{H}^n$ under $\mathrm{CD}(\pm 1, m)$ for $m \geq n$ respectively.
- (Jacobi polynomials: $y = \mathscr{P}^{(\alpha,\beta)}_k(t)$ with $\alpha, \beta > -1, k \geq 0$) Here $\mathscr{M} = (-1, 1)$ and $d\mu = (1-t)^\alpha(1+t)^\beta dt$,

$$\left(1-t^2\right)\frac{d^2y}{dt^2} - (\alpha - \beta + (\alpha+\beta+2)t)\frac{dy}{dt} + k(k+\alpha+\beta+1)y = 0.$$

- Orthogonal Polynomials (Bessel, Hermite, Jacobi, Laguerre, Tchebychev, Ultra-spherical, Legendre, Hypergeometric series, etc.) and Sturm-Liouville systems.

The curvature-dimension condition $\mathrm{CD}(k, m)$. In the context of smooth metric measure spaces one imposes Ricci curvature lower bounds in the form $\mathscr{R}ic_\varphi(g) \geq kg$ or the stronger form $\mathscr{R}ic^m_\varphi(g) \geq kg$ for some fixed constants $k \in \mathbb{R}$ and $n \leq m < \infty$. This enables one to use various comparison results (e.g., Hessian and Laplace comparison, Bishop-Gromov volume comparison, *etc.*) that are crucial instruments in doing analysis and geometry.

A curvature lower bound in the form $\mathscr{R}ic^m_\varphi(g) \geq kg$ (with $k \in \mathbb{R}$) implies that the diffusion operator $L = \Delta_\varphi$ satisfies the curvature-dimension condition $\mathrm{CD}(k, m)$. To elaborate on this last point further, we first recall the definition of the carre du champ operator $\Gamma[L]$ associated with a Markov diffusion operator L. This is given by

$$\Gamma[L](u, v) = \Gamma_1[L](u, v) := \frac{1}{2}[L(uv) - uLv - vLu]. \tag{12}$$

Higher order iterates are then defined inductively by replacing the product operation with Γ respectively. Indeed, the second order iterated carre du champ operator $\Gamma_2[L]$, can be seen to be,

$$\Gamma_2[L](u, v) := \frac{1}{2}[L\Gamma(u, v) - \Gamma(u, Lv) - \Gamma(Lu, v)]. \tag{13}$$

For $u = v$ we write $\Gamma[L](u) = \Gamma[L](u, u)$ and $\Gamma_2[L](u) = \Gamma_2[L](u, u)$. Now the Markov diffusion operator L is said to satisfy the Bakry-Émery curvature-dimension condition $\mathrm{CD}(k, m)$ *iff* for some $m \geq 2$ and $k \in \mathbb{R}$,

$$\Gamma_2[L](w) = \frac{1}{2}L\Gamma(w) - \Gamma(w, Lw) \geq \frac{1}{m}(Lw)^2 + k\Gamma[L](w). \tag{14}$$

By a direct calculation, it is seen that for $L = \Delta_\varphi$ as in (7) and $\Gamma[L]$ and $\Gamma_2[L]$ as in (12) and (13), we have the first and second order relations

$$\Gamma[L](w) = |\nabla w|^2, \tag{15}$$

$$\Gamma_2[L](w) = \frac{1}{2}L|\nabla w|^2 - \langle \nabla w, \nabla Lw\rangle, \tag{16}$$

respectively. Hence, in light of (11), (15) and (16) it follows from the curvature lower bound $\mathscr{R}ic^m_\varphi(g) \geq kg$ that here the iterated carre du champ operator $\Gamma_2[L]$ satisfies the inequality

$$\begin{aligned}\Gamma_2[L](w) &\geq \frac{1}{m}(Lw)^2 + \mathscr{R}ic^m_\varphi(\nabla w, \nabla w)\\ &\geq \frac{1}{m}(Lw)^2 + k|\nabla w|^2 = \frac{1}{m}(Lw)^2 + k\Gamma[L](w).\end{aligned}$$

In contrast, a curvature lower bound in the form $\mathscr{R}ic_\varphi(g) \geq kg$ implies the curvature-dimension condition $\mathrm{CD}(k, \infty)$, in the sense that by virtue of (10), (15) and (16) one only has

$$\Gamma_2[L](w) = |\nabla\nabla w|^2 + \mathscr{R}ic_\varphi(\nabla w, \nabla w) \geq k|\nabla w|^2 = k\Gamma[L](w). \tag{17}$$

Curvature lower bounds and/or curvature-dimension conditions are instrumental for doing analysis and geometry on smooth metric measure spaces. Below we see their significance in Poincaré and logarithmic Sobolev inequalities. First let us look at some equivalent versions of the conditions formulated in terms of the semigroup $P_t = e^{t\Delta_\varphi}$. Below we assume that $d\mu$ is a probability measure and throughout we take $f > 0$.

Theorem 1 *The following conditions for $k > 0$ are equivalent:*

- ▷ (*Curvature-Dimension*) $\mathrm{CD}(k, \infty)$.[3]
- ▷ (*Commutation*) $|\nabla P_t f| \leq e^{-kt} P_t |\nabla f|$.
- ▷ (*Contraction in Wasserstein Space*)

$$W_2(P_t f d\mu, P_t g d\mu) \leq e^{-kt} W_2(f d\mu, g d\mu).$$

Theorem 2 $\mathrm{CD}(k, \infty) \implies$ *Poincaré inequality with constant* $1/k$.

Proof Note that Poincaré inequality (spectral gap) amounts to showing the inequality $\mathrm{Var}_\mu(f) \leq (1/\lambda)\mathrm{E}_\mu(f)$ where the smallest such $\lambda > 0$ is the first (non-zero) eigenvalue of $-\Delta_\varphi$ on $\mathscr{M}$. Thus we write

$$\begin{aligned}
\mathrm{Var}_\mu(f) &= \int_{\mathscr{M}} f^2 \, d\mu - \left(\int_{\mathscr{M}} f \, d\mu\right)^2 = \int_{\mathscr{M}} \left| f - \int_{\mathscr{M}} f \, d\mu \right|^2 d\mu \\
&= -\int_0^\infty \frac{d}{dt} \int_{\mathscr{M}} [P_t f]^2 \, d\mu \, dt = -2 \int_0^\infty \int_{\mathscr{M}} [\Delta_\varphi P_t f][P_t f] \, d\mu \, dt \\
&= 2 \int_0^\infty \int_{\mathscr{M}} |\nabla P_t f|^2 \, d\mu \, dt \leq 2 \int_0^\infty \int_{\mathscr{M}} e^{-2kt} P_t |\nabla f|^2 \, d\mu \, dt \\
&\leq \frac{1}{k} \int_{\mathscr{M}} |\nabla f|^2 \, d\mu = \frac{1}{k} \mathrm{E}_\mu(f)
\end{aligned}$$

as required. □

[3] See (14)–(17), in particular, the equivalence with $\mathscr{R}ic_\varphi(g) \geq kg$. On $\mathbb{R}^n$ this is the *uniform convexity* of φ, i.e., $\mathscr{R}ic_\varphi(g) \geq kg \iff \nabla\nabla\varphi \geq k\mathrm{I}$.

Remark 1 The Poincaré inequality $\mathrm{Var}_\mu(f) \le (1/k)\mathrm{E}_\mu(f)$ gives exponential decay in *variance*: $\mathrm{Var}_\mu(P_t f) \le e^{-2kt}\mathrm{Var}_\mu(f)$. Note that the curvature-dimension condition CD(k, m) gives a Poincaré inequality with constant $(m-1)/(mk)$.

Theorem 3 CD(k,∞) $\Longrightarrow$ *Logarithmic Sobolev inequality with constant* $1/k$.

Proof The claim amounts to showing $\mathrm{Ent}_\mu(f^2) \le 1/(2k) I_\mu(f^2) = (2/k)\mathrm{E}_\mu(f)$. Thus we write

$$\begin{aligned}
\mathrm{Ent}_\mu(f) &= \int_{\mathscr{M}} f \log f \, d\mu - \int_{\mathscr{M}} f \, d\mu \log \int_{\mathscr{M}} f \, d\mu \\
&= -\int_0^\infty \frac{d}{dt} \int_{\mathscr{M}} [P_t f \log P_t f] \, dt \, d\mu = -\int_0^\infty \int_{\mathscr{M}} [\Delta_\varphi P_t f][\log P_t f] \, dt \, d\mu \\
&= \int_{\mathscr{M}} \int_0^\infty \frac{|\nabla P_t f|^2}{P_t f} \, dt \, d\mu \le \int_{\mathscr{M}} \int_0^\infty e^{-2kt} \frac{[P_t |\nabla f|]^2}{P_t f} \, dt \, d\mu.
\end{aligned}$$

Now by Cauchy-Schwarz inequality

$$[P_t|\nabla f|]^2 = \left[P_t\left(\sqrt{f}\frac{|\nabla f|}{\sqrt{f}}\right)\right]^2 \le P_t f \times P_t\left[\frac{|\nabla f|^2}{f}\right].$$

This yields

$$\begin{aligned}
\mathrm{Ent}_\mu(f) &= \int_{\mathscr{M}} f \log f \, d\mu - \int_{\mathscr{M}} f \, d\mu \log \int_{\mathscr{M}} f \, d\mu \\
&\le \int_{\mathscr{M}} \int_0^\infty e^{-2kt} P_t\left[\frac{|\nabla f|^2}{f}\right] dt \, d\mu \le \frac{1}{2k} \int_{\mathscr{M}} \frac{|\nabla f|^2}{f} \, d\mu = \frac{1}{2k} I_\mu(f),
\end{aligned} \tag{18}$$

as required. □

Remark 2 Logarithmic Sobolev inequatiy (18) gives exponential decay in entropy: $\mathrm{Ent}_\mu(P_t f) \le e^{-2kt}\mathrm{Ent}_\mu(f)$.

Theorem 4 CD(k,∞) $\Longrightarrow$ *Isoperimetric inequality* $\sqrt{k} I(\mu(A)) \le \mu^+(A)$ *where* I *is the isoperimetric profile of the Gaussian* [*as in* (3)].

5 Nonlinear Diffusion Equations on Smooth Metric Measure Spaces

In this last section we look more closely at (nonlinear) diffusion equations on smooth metric measure spaces. Here we confine to the porous medium and the fast diffusion equations ([16]–[17]).[4] For a (space-time) function $u = u(x,t) > 0$ and fixed exponent $p > 0$ this can be written in the form

$$\Box_p u(x,t) = \partial_t u(x,t) - \Delta_\varphi u^p(x,t) = \mathscr{N}(t,x,u(x,t)). \tag{19}$$

Here $\mathscr{N} = \mathscr{N}(t,x,u)$ is a sufficiently smooth nonlinearity depending on the space-time variable (x,t) and the dependent variable (solution) u. The porous medium equation corresponds to the range $p > 1$ and the fast diffusion equation to the range $0 < p < 1$.

We write $\mathscr{B}_R(x_0)$ for the closed geodesic ball in $\mathscr{M}$ centered at x_0 with radius $R > 0$. Likewise we write $Q_{R,T}(x_0)$ for the compact parabolic space-time cylinder with lower base $\mathscr{B}_R(x_0) \times \{t_0 - T\}$ for $t_0 \in \mathbb{R}$ and height $T > 0$, specifically, $Q_{R,T}(x_0) = \mathscr{B}_R(x_0) \times [t_0 - T, t_0] \subset \mathscr{M} \times (-\infty, \infty)$. When the choice (x_0, t_0) is clear from the context we write $\mathscr{B}_R$, $Q_{R,T}$. We write $X_+ = \max(X, 0)$ and $X_- = \max(-X, 0)$.

The nonlinear porous medium equation. The first nonlinear diffusion equation we look at here is the one of slow diffusion type (or equivalently porous medium type). This is (19) with $p > 1$. We also set

$$\Sigma(t,x,v) = p[(p-1)v/p]^{\frac{p-2}{p-1}} \mathscr{N}\left(t, x, [(p-1)v/p]^{\frac{1}{p-1}}\right).$$

Theorem 5 *Let $(\mathscr{M}, g, d\mu)$ be a complete smooth metric measure space with $d\mu = e^{-\varphi} dv_g$ and assume $\mathscr{R}ic_\varphi^m(g) \geq -(m-1)kg$ in $\mathscr{B}_R$ for some $k \geq 0$, $m \geq n$ and $R > 0$. Let u be a positive solution to* (19) *with $1 < p < 1 + 1/\sqrt{m-1}$ and $v = pu^{p-1}/(p-1)$ and $M = \sup_{Q_{R,T}} v$. Then there exists $C = C(p,m) > 0$ such that for every (x,t) in $Q_{R/2,T}$ with $t > t_0 - T$ we have*

$$v^{\frac{1}{2(p-1)}} |\nabla v|(x,t) \leq C \left\{ \begin{array}{l} \left[\dfrac{k^{1/4}}{\sqrt{R}} + \dfrac{1}{R} + \sqrt{k}\right] M^{1+\frac{1}{2(p-1)}} \\ + \sup_{Q_{R,T}} \left\{ v^{p/[2(p-1)]} \left[2\Sigma_v(t,x,v) + \dfrac{\Sigma(t,x,v)}{(p-1)v}\right]_+^{1/2} \right\} \\ + \dfrac{M^{\frac{p}{2(p-1)}}}{\sqrt{t - t_0 + T}} + \sup_{Q_{R,T}} \left\{ \left[v^{(2p+1)/[2(p-1)]} |\Sigma_x(t,x,v)|\right]^{1/3} \right\} \end{array} \right\}.$$

[4] Diffusion equations have a close affinity with geometric and functional inequalities. One strong connection comes from the study of entropies and gradient flows in Wasserstein spaces. In a related paper in this volume (by V. Vahidifar) the nonlinear heat equation $\Box u(x,t) = \partial_t u(x,t) - \Delta_\varphi u(x,t) = \mathscr{N}(t,x,u(x,t))$ and their gradient estimates on smooth metric measure spaces are discussed.

Theorem 6 *Let $(\mathscr{M}, g, d\mu)$ be a complete smooth metric measure space with $d\mu = e^{-\varphi} dv_g$ and $\mathscr{R}ic^m_\varphi(g) \geq 0$. Assume $(3-2p)\mathscr{N}(u) - 2u\mathscr{N}_u(u) \geq 0$ for all $u > 0$ where $1 < p < 1 + 1/\sqrt{m-1}$. Then any positive ancient solution to the nonlinear porous medium equation*

$$\Box_p u(x,t) = \partial_t u(x,t) - \Delta_\varphi u^p(x,t) = \mathscr{N}(u(x,t)), \tag{20}$$

satisfying the growth at infinity $u(x,t) = o([\varrho(x) + \sqrt{|t|}]^{2/(2p-1)})$, must be spatially constant. In particular, if additionally, $\mathscr{N}(u) \geq a$ for some $a > 0$ and all $u > 0$ then (20) *admits no such ancient solutions.*

Theorem 7 *Let $(\mathscr{M}, g, d\mu)$ be a smooth metric measure space with $\mathscr{M}$ closed and $d\mu = e^{-\varphi} dv_g$ and let $\mathscr{R}ic^m_\varphi(g) \geq -(m-1)kg$ on $\mathscr{M}$ for some $k \geq 0$. Let u be a positive smooth solution to* (19) *where $1 < p \leq 1 + 1/\sqrt{m-1}$ and set $v = pu^{p-1}/(p-1)$. Let $\Gamma = \Gamma(v)$ with $v > 0$ be a given function of class $\mathscr{C}^2$ and assume the following conditions hold for some a:*

- $\Gamma'(v)\Sigma(t,x,v) \leq 0$,
- $\Gamma'(v) + v\Gamma''(v) \geq 0$,
- $\langle \nabla v, \Sigma_x(t,x,v)\rangle \leq 0$,
- $\Sigma_v(t,x,v) + \Sigma(t,x,v)/[2(p-1)v] \leq a$.

Then for every $x \in \mathscr{M}$ and $0 < t \leq T$ we have the gradient bound

$$[v^{\frac{1}{p-1}}|\nabla v|^2](x,t) \leq e^{2(MK+a)t} \left\{ \max_{\mathscr{M}} \left[v^{\frac{1}{p-1}}|\nabla v|^2 + \Gamma(v) \right]_{t=0} - \Gamma(v(x,t)) \right\},$$

where $M = \sup v$ and $K = (p-1)(m-1)k$.

Theorem 8 *Let $(\mathscr{M}, g, d\mu)$ be a smooth metric measure space with $\mathscr{M}$ closed and $d\mu = e^{-\varphi} dv_g$ and let $\mathscr{R}ic^m_\varphi(g) \geq -(m-1)kg$ on $\mathscr{M}$ for some $k \geq 0$. Let u be a positive smooth solution to* (19) *with $1 < p \leq 1 + 1/\sqrt{m-1}$ and set $v = pu^{p-1}/(p-1)$. Assume the following conditions hold:*

- $\Sigma(t,x,v) \leq 0$,
- $\langle \nabla v, \Sigma_x(t,x,v)\rangle \leq 0$,
- $\Sigma_v(t,x,v) + \Sigma(t,x,v)/[2(p-1)v] \leq 0$.

Then for every $x \in \mathscr{M}$ and $0 < t \leq T$ we have the gradient bound

$$\frac{p^2 t}{1 + 2MKt}[v^{\frac{1}{p-1}}|\nabla v|^2](x,t) \leq (p-1)\left[\max_{\mathscr{M}} v^{\frac{p}{p-1}}(x,0) - v^{\frac{p}{p-1}}(x,t) \right],$$

where $M = \sup v$ and $K = (p-1)(m-1)k$.

The nonlinear fast diffusion equation. This is (19) with $0 < p < 1$. For technical reasons we introduce $p_0 = p_0(m) > 0$ as the larger of the values $1/2$ and $1 - 1/\sqrt{m-1}$ $(m \geq 2)$, that is, $p_0 = 1/2$ when $2 \leq m \leq 5$ and $p_0 = 1 - 1/\sqrt{m-1}$ when $m \geq 5$. Thus we always have $p_0(m) \geq 1/2$, and when $m \geq 4$ we have $p_0 \leq p_c = 1 - 2/m$. We also set

$$\Sigma^\star(t, x, v) = (p - 1/2) v^{1 - \frac{1}{p-1/2}} \mathscr{N}(t, x, v^{\frac{1}{p-1/2}}).$$

Theorem 9 *Let $(\mathscr{M}, g, d\mu)$ be a complete smooth metric measure space with $d\mu = e^{-\varphi} dv_g$ and $\mathscr{R}ic_\varphi^m(g) \geq -(m-1)kg$ in $\mathscr{B}_R$ for some $k \geq 0$, $m \geq n$ and $R > 0$. Let u be a positive solution to (19) with $p_0 < p < 1$, $v = u^{p-1/2}$ and $M = \sup_{Q_{R,T}} v$. Then there exists $C = C(p, m) > 0$ such that for (x, t) in $Q_{R/2,T}$ with $t > t_0 - T$ we have*

$$|\nabla v|(x, t) \leq C \left\{ \begin{array}{l} \dfrac{M^{\frac{p}{2p-1}}}{\sqrt{t - t_0 + T}} + \sup_{Q_{R,T}} \left\{ v^{\frac{2p}{3(2p-1)}} |\Sigma_x^\star(t, x, v)|^{1/3} \right\} \\ + \left[\dfrac{k^{1/4}}{\sqrt{R}} + \dfrac{1}{R} + \sqrt{k} \right] M + \sup_{Q_{R,T}} \left\{ v^{\frac{p}{2p-1}} [\Sigma_v^\star(t, x, v)]_+^{1/2} \right\} \end{array} \right\}.$$

Theorem 10 *Let $(\mathscr{M}, g, d\mu)$ be a complete smooth metric measure space with $d\mu = e^{-\varphi} dv_g$ and $\mathscr{R}ic_\varphi^m(g) \geq 0$. Assume $(3 - 2p)\mathscr{N}(u)/u - 2\mathscr{N}_u(u) \geq 0$ for all $u > 0$ where $p_0 < p < 1$. Then any positive ancient solution to the nonlinear fast diffusion equation*

$$\Box_p u(x, t) = \partial_t u(x, t) - \Delta_\varphi u^p(x, t) = \mathscr{N}(u(x, t)), \tag{21}$$

satisfying the growth at infinity $u(x, t) = o([\varrho(x) + \sqrt{|t|}]^{2/p})$ must be spatially constant. If, in addition, $\mathscr{N}(u) \geq a$ for some $a > 0$ and all $u > 0$ then (21) admits no such ancient solutions.

Theorem 11 *Let $(\mathscr{M}, g, d\mu)$ be a smooth metric measure space with $\mathscr{M}$ closed and $d\mu = e^{-\varphi} dv_g$ and let $\mathscr{R}ic_\varphi^m(g) \geq -(m-1)kg$ on $\mathscr{M}$ for some $k \geq 0$. Let u be a positive smooth solution to (19) where $p_0 < p < 1$ and set $v = u^{p-1/2}$. Let $\Gamma = \Gamma(v)$ with $v > 0$ be a given function of class $\mathscr{C}^2$ and assume the following conditions hold for some a:*

- $\Sigma^\star(t, x, v) \leq a$,
- $\Gamma'(v) \Sigma^\star(t, x, v) \leq 0$,
- $\Gamma'(v) + v\Gamma''(v) \geq 0$,
- $\langle \nabla v, \Sigma_x^\star(t, x, v) \rangle \leq 0$.

Then for all $x \in \mathscr{M}$ and $0 < t \leq T$ we have

$$|\nabla v|^2(x, t) \leq e^{2(MK+a)t} \left\{ \max_{\mathscr{M}} \left[|\nabla v|^2 + \Gamma(v) \right]_{t=0} - \Gamma(v(x, t)) \right\},$$

where $M = \sup v$ and $K = (p-1)(m-1)k$.

Theorem 12 *Let $(\mathscr{M}, g, d\mu)$ be a smooth metric measure space with $\mathscr{M}$ closed and $d\mu = e^{-\varphi} dv_g$ and let $\mathscr{R}ic^m_\varphi(g) \ge -(m-1)kg$ on $\mathscr{M}$ for some $k \ge 0$. Let u be a positive smooth solution to* (19) *where $p_0 < p < 1$ and set $v = u^{p-1/2}$. Assume the following conditions hold:*

- $\Sigma^\star(t, x, v) \le 0$,
- $\Sigma^\star_v(t, x, v) \le 0$,
- $\langle \nabla v, \Sigma^\star_x(t, x, v) \rangle \le 0$.

Then for $x \in \mathscr{M}$ and $0 < t \le T$ we have

$$\frac{t|\nabla v(x,t)|^2}{1+2MKt} \le \frac{(2p-1)^2}{8p^3} \left[\max_{\mathscr{M}} v^{\frac{2p}{2p-1}}(x,0) - v^{\frac{2p}{2p-1}}(x,t) \right],$$

where $M = \sup v$ and $K = (p-1)(m-1)k$.

Acknowledgements I wish to express my sincere gratitude to Prof Ruzhansky for the opportunity to visit the Ghent Analysis and PDE centre during Summer 2024. I also wish to thank the organisers of the workshop "Inverse Problems and Application" for organising such a nice, timely and stimulating meeting. Support from the Engineering and Physical Sciences Research Council (EPSRC) through the grant EP/V027115/1 is gratefully acknowledged.

References

1. Ambrosio, L., Gigli, N., Savaré, G.: Gradient flows: In: Metric Spaces and the Space of Probability Measures, Lectures in Mathematics, ETH Zürich, 2nd Edition, Birkhäuser (2008)
2. Ambrosio, L., Mondino, A., Savaré, G.: Nonlinear diffusion equations and curvature conditions in metric measure spaces, memoirs American mathematical society. 1270, AMS (2019)
3. Bakry, D., Gentil, I., Ledoux, M.: Analysis and geometry of markov diffusion operators, a series of comprehensive studies in mathematics. 348, Springer (2012)
4. Cavaletti, F., Mondino, A.: Sharp geometric and functional inequalities in metric measure spaces with lower Ricci curvature bounds. Geom. Top. **21**, 603–645 (2017)
5. Cavaletti, F., Mondino, A.: Sharp and rigid isoperimetric inequalities in metric measure spaces with lower Ricci curvature bounds. Invent. Math. **208**, 803–849 (2017)
6. Milman, E.: Sharp isoperimetric inequalities and model spaces fpr the Curvature-Dimension-Diameter condition. J. Eur. Math. Soc. **17**, 1041–1078 (2015)
7. Otto, F.: The geometry of dissipative evolution equations: The porous medium equation. Comm. Partial Differ. Equ. **26**(1–2), 101–174 (2001)
8. Ruzhansky, M., Turunen, V.: Pseudo differential operators and symmetries: Background analysis and advanced topics, Birkhäuser (2010)
9. Ruzhansky, M., Suragan, D.: Hardy inequalities on homogeneous groups: 100 years of Hardy inequalities, progress in mathematics, vol **2**, Birkhäuser (2019)
10. Taheri, A.: Function spaces and partial differential equations, vol. **I**, Oxford Lecture Series in Mathematics and its Applications, **40**, OUP (2015)
11. Taheri, A.: Function spaces and partial differential equations, vol. **II**, Oxford Lecture Series in Mathematics and its Applications **41**, OUP (2015)

12. Taheri, A.: Liouville theorems and elliptic gradient estimates for a nonlinear parabolic equation involving the Witten Laplacian, Published online in: Adv. Calc. Var., De Gruyter (2021)
13. Taheri, A.: Gradient estimates for a weighted Γ-nonlinear parabolic equation coupled with a super Perelman-Ricci flow and implications, Published online in: Potential Anal., Springer (2021)
14. Taheri, A., Vahidifar, V.: Gradient estimates for a nonlinear parabolic equation on smooth metric measure spaces with evolving metrics and potentials, Nonlinear Anal., 232 (2023)
15. Taheri, A., Vahidifar, V.: Curvature conditions, Liouville-type theorems and Harnack inequalities for a nonlinear parabolic equation on smooth metric measure spaces. De Gruyter, Adv. Nonlinear Studies (2024)
16. Taheri, A., Vahidifar, V.: The nonlinear porous medium equation for the f-Laplacian: Hamilton-Souplet-Zhang type gradient estimates and implications, Submitted for publication (2024)
17. Taheri, A., Vahidifar, V.: The nonlinear fast diffusion equation on smooth metric measure spaces: Hamilton-Souplet-Zhang estimates and a Ricci Perelman super flow. Calc. Var. & PDEs **64**, 1–36 (2025)
18. Villani, C.: Optimal transport: Old and New, A Series of comprehensive studies in mathematics, **338**, Springer (2008)
19. Wang, F.Y.: Analysis for Diffusion Processes on Riemannian Manifolds, Advanced Series on Statistical Science & Probability, Vol. **18**, World Scientific (2013)
20. Zhang, Q.S.: Sobolev inequalities, heat kernels under Ricci flow and the Poincaré conjecture, CRC Press (2011)

The Historical Background of A. Markov's Inequality

Armen Vagharshakyan

Abstract A. Markov's inequality was the first ever inequality proved for derivatives of polynomials: given a polynomial of a particular degree and a bound on its maximum on an interval, it provides the sharp estimate on the maximum of its derivative on that interval. A. Markov proved it to answer a question posed by D. Mendeleev regarding alcohol measurements. We briefly present that story.

Keywords Ethanol-water mixture · Inequalities for derivatives of polynomials · Markov's inequality

MSC2020: 26D05 · 41A17 · 26D10

1 The Paper Structure

The paper is structured as follows. Section 2 describes the historical importance of alcohol measurements in general. Section 3 presents the original motivation of A. Markov to answer a question posed by D. Mendeleev regarding alcohol measurements. Section 4 explains the underlying chemical phenomena explaining D. Mendeleev's observations. Finally, Sect. 5 contains further references as well as some minor corrections to the existing literature.

A. Vagharshakyan (✉)
Institute of Mathematics NAS RA, Yerevan, Armenia

Yerevan State University, Yerevan, Armenia
e-mail: avaghars@kent.edu

M. Chatzakou et al. (eds.), *Direct and Inverse Problems with Applications*,
Trends in Mathematics 12, https://doi.org/10.1007/978-3-031-98645-1_22

2 Alcohol Measurements

As pointed out in [1]: "The first taxes on alcohol in Europe date back to Antiquity. The Romans, for example, levied a tax on the sale of wine. During the Middle Ages, alcohol taxes were often used to finance wars and crusades. Alcohol taxes were a particularly attractive source of revenue because they were easy to collect and brought in a lot of money."

For instance, as noted in [5]: "A substantial part of the Dutch taxation system consisted of taxes on beer. In fact, the excise tax on beer was one of the largest components of government revenues in Holland, the leading province in the Dutch Republic. Hence, beer taxes played a crucial role in giving the Dutch Republic the military power to break away from the Spanish-occupied Low Countries and thus in the determination of the present-day border between Belgium and the Netherlands."

However, as also noted in [5]: "Fraud was commonplace as brewers tried to circumvent the beer taxes by all means."

This is why, for the purpose of effective taxation, governments of various countries started requesting scientists to create reference tables for determining ethanol percentage in alcoholic beverages, based on their relative density (i.e. the density with respect to pure water's density at a specific temperature), also referred to as 'specific gravity' later in this article.

For instance, as noted in [2]: "On June 10, 1857, the Minister of Finance (of Netherlands), having consulted the Academy on the complaints of some distillers regarding the method of taxing the liquor, has appointed on this subject a commission relating to the liquor-scale tables" (translated from French, slightly abridged).

A similar request to scientists has been made earlier by the government of Great Britain. As a result, in 1792, G. Gilpin performed a series of experiments and, together with Ch. Blagden, wrote the "Supplementary Report on the best Method of proportioning the Excise upon Spirituous Liquors" and "Tables for reducing the Quantities by Weight, in any Mixture of Pure Spirit and Water" (see [3],[6]). We refer the reader to the article [24], published by the Captain Cook Society, for this and for the otherwise fascinating biography of G. Gilpin.

In turn, the Ministry of Finances of the Russian Empire has tasked D. Mendeleev with refining the tables for the specific gravity of various types of wines and alcohols (see [7]). D. Mendeleev conducted relevant experiments while visiting Hottingen in 1859. His calculations were so neat that they weren't surpassed until modern times. In 1865, D. Mendeleev presented them in his doctoral thesis [8]. He has also served as an expert on chemical technologies in the works of Russia's government commission on excise taxes [9]. In 1887, D. Mendeleev published the manuscript [10] "A research on aqueous solutions by their specific gravity" where his twenty-year expertise on the subject has been compiled. We will elaborate on the manuscript in the next section.

3 A Question by D. Mendeleev

The story of the manuscript [10] is remarkable on its own. Having been originally published in 1887, it was lost, and a copy of it was found in D. Mendeleev's personal archives much later by E. Fritzman, who was the head of Mendeleev's museum from 1931 up to his death in 1942, during the siege of Leningrad. The manuscript was finally published in 1934 as a part of D. Mendeleev's selected works (while being censored for some of D. Mendeleev's religious observations).

In its introduction (see [10], p. XIII) it is stated that originally he only planned to present a compilation of reference tables containing experimental results on the dependence of the resulting density of the water solution of ethanol (or some other soluble organic compound like salt, sulfuric acid, acetic acid) on the percentage of ethanol by mass, at a given temperature.

However, as D. Mendeleev was compiling all credible results ever obtained, he noticed a circumstance which made him review the whole work and start it anew–this time from a theoretical point of view. Let us explain it now.

D. Mendeleev plotted the experimental results as points on a plane and then utilized the method of least squares to find best fitting polynomials. Remarkably, while applying the method of least squares, he was consulted by his fellow mathematicians, including P. Chebyshev (see [8], p. 106).

As a result, D. Mendeleev noted the following circumstance. The range of alcohol percentages naturally split into four intervals. At each of the intervals, the best fitting resulted in a polynomial of order two, in the sense that its third-order and higher-order terms had negligible coefficients. From that, he concluded that, in reality, there are just four patterns, each described by a polynomial of order two (i.e., a parabola), with all deviations being attributed to mistakes in experiments. In Fig. 1, one can see the four linear segments that are the graphs of the derivatives of the mentioned parabolas, as plotted by D. Mendeleev himself (see [10], p. 255). The points where the pattern changes from one linear segment to another roughly correspond to the percentage of alcohol (by mass) in wine, vodka and absinthe.

Then, the problem that intrigued D. Mendeleev was whether the jump discontinuities in Fig. 1 were due to experimental errors or were genuine (i.e., they were supposed to be there). In order to answer that problem, based on his chemical observations, D. Mendeleev first evaluated the maximal possible error L in experiments. The mathematical question was then: how much did an error of L in experiments affect the derivative of the parabola? That is, for a polynomial f of order $n = 2$, assume that

$$|f(z)| \leq L, \quad a \leq z \leq b$$

What is then the best bound for the derivative $f'(z)$?

D. Mendeleev was only able to solve the question under such additional assumptions that the function f could be determined uniquely (see [10], p. 256), for instance:

$$f(a) = -L, \ f\left(\frac{a+b}{2}\right) = 0, \ f(b) = L.$$

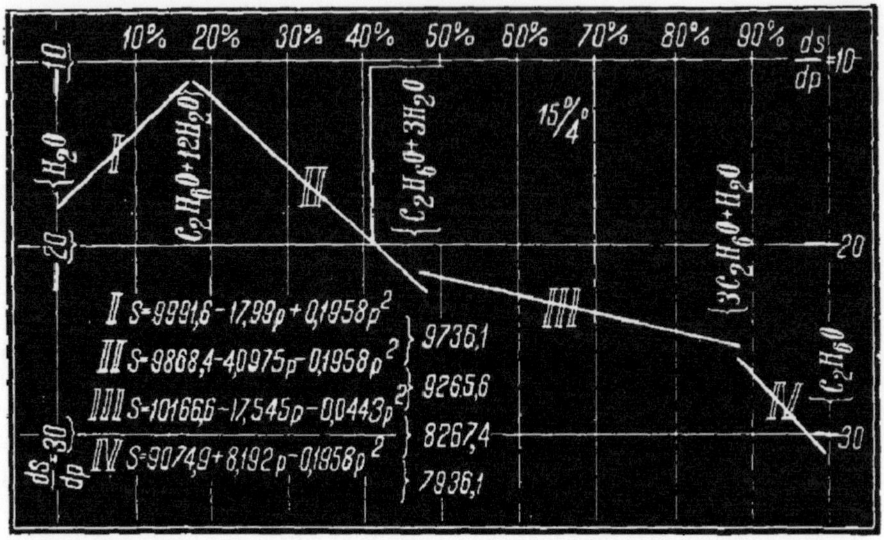

Fig. 1 The derivative of solution's density with respect to alcohol percentage (by mass) [10]

After learning that question, A. Markov answered it in full generality. That is, in his article "On a question by D.I. Mendeleev" [11], A. Markov proposed and solved the following problem.

Consider a polynomial f of order at most n. Assume that

$$|f(z)| \leq L, \quad a \leq z \leq b.$$

What is then the best bound for the derivative $f'(z)$?

A. Markov's solution is the first ever inequality proved for derivatives of polynomials. Namely, in [11] he proved that

$$|f'(z)| \leq \frac{2n^2 L}{b-a}, \quad a \leq z \leq b.$$

Regretfully, the value L that D. Mendeleev had obtained was not neat enough to answer the problem of jump discontinuities in a definite way (see [10], p. 259). Consequently, as he himself phrased (see [10], p. 221): "After working for some twenty-two years on the issue, I was not resolute to state a final judgement on the question, but I don't want to further delay the bulk of the observations that I have collected, as I find the collected data very important for theory and practise ... but life goes on, and one has to realise that I myself won't be able to fulfil everything that I want to, and, consequently, I must publish my unfinished work in order to ease the work of others" (translated from Russian, edited).

The author does not know if the answer to D. Mendeleev's question has been obtained so far. However, modern knowledge sheds light to the circumstance observed by D. Mendeleev. It is briefly discussed in the next section.

4 Hydrogen Bonds

The reason that the density of ethanol solution in water doesn't increase linearly with respect to the ethanol percentage, as one would naively expect, is the existence of forces of atomic attraction and distraction and, specifically, those that cause hydrogen bonds: a phenomena that would only be hypothesised by D. Mendeleev and that would be discovered much later, in 1912. We refer the reader to [12] for the history of discovering hydrogen bonds.

We now describe how these forces affect the solution's density in an exceedingly schematic and oversimplified way. As ethanol is being added to pure water, an individual ethanol molecule starts changing the spatial pattern of water molecules due to forces of attraction and distraction in between the atoms of water H_2O and the atoms of ethanol C_2H_5OH. Strong $O - H$ and weak $C - H$ hydrogen bonds (in between corresponding atoms of different molecules) are prime examples of such forces (see e.g. [13] for a discussion of the resulting ethanol-water dimer structure due to hydrogen bonds). As a result, water molecules redistribute in space so as to form a type 1–12 macromolecule (or, to be precise, a cage-like structure called macromolecular cage) consisting of a single ethanol molecule surrounded by some 12 water molecules. As ethanol molecules keep being added to the solution, more type 1–12 macromolecules are being formed. Correspondingly, the density of the solution changes according to the parabolic law I from Fig. 1. Formation of such macromolecules continues until the ratio of ethanol molecules to water molecules in the solution reaches 1 to 12 (as mentioned in Fig. 1).

After that stage, if still more ethanol molecules are added to the solution, the situation changes drastically. Indeed, new ethanol molecules cannot form type 1–12 macromolecules anymore, since each water molecule in the solution is already a part of some preexisting type 1–12 macromolecule. As a consequence, macromolecules of a new type 1–3, consisting of a single ethanol molecule surrounded by three water molecules, are being formed. The spatial distribution and the density of type 1–3 macromolecules differs from those of type 1–12 macromolecules. Correspondingly, the density of the solution starts changing according to a different parabolic law: this time, law II from Fig. 1.

Of course, this is an oversimplified explanation. In fact, as noted in [14]: "Unfortunately, the picture we possess nowadays about this essential property is somewhat confusing and contradictory."

For a more elaborate insight into the phenomena, we refer the reader to the article [15] from "Nature" whose abstract states: "When a simple alcohol such as methanol or ethanol is mixed with water, the entropy of the system increases far less than expected for an ideal solution of randomly mixed molecules. This well-known effect has been attributed to hydrophobic headgroups creating ice-like or clathrate-like structures in the surrounding water, although experimental support for this hypothesis is scarce. In fact, an increasing amount of experimental and theoretical work suggests that the hydrophobic headgroups of alcohol molecules in aqueous solution cluster together... Our data indicate that most of the water molecules exist as small hydrogen-bonded strings and clusters in a 'fluid' of close-packed methyl groups, with water clusters bridging neighbouring methanol hydroxyl groups through hydrogen bonding. This behaviour suggests that the anomalous thermodynamics of water alcohol systems arises from incomplete mixing at the molecular level and from retention of remnants of the three-dimensional hydrogen-bonded network structure of bulk water."

5 Further Reading

For further reading on the general history of inequalities for the derivatives of polynomials, we refer the reader to the surveys of R. Boas [16] and B. Saffari [17]. We point out some minor corrections in these two articles: as we have already mentioned, D. Mendeleev didn't prove the case $n = 2$ in full generality (unlike the claim in [17]); the fact that the four parabolas mentioned in Fig. 1 are different was apparent (unlike the claim in [16]), instead, the existence of jump discontinuities of the derivatives was in question.

A compilation of proofs of A. Markov's inequality and an extensive list of literature can be found in a survey of A. Shadrin [18].

Further generalizations and extensions of A. Markov's inequality include, among many others, its generalization for higher-order derivatives by his brother V. Markov [19], the results of P. Erdös [20] and P. Borwein [21] for polynomials with real roots, the result of S. Konyagin [22] on Markov's inequality in L_1, a list of works of A. Shadrin and his coauthors (see e.g [23]) on Markov's inequality in weighted spaces, the result of D. Wilhelmsen [25] on multidimensional analogues, the result of M. Chatzakou and Y. Sarantopoulos [4], and the references therein, for polynomials on Banach spaces.

In turn, for more research on ethanol-water packings and resulting macromolecules, we refer the reader to: the above-mentioned articles [13], [14], [15], the article [26], whose Fig. 2 may serve as a modern analogue of Fig. 1, and the references therein.

Acknowledgements This work was supported by the Higher Education and Science Committee of the Republic of Armenia (Research Project No. 24RL-1A028). The review itself is based on the author's talk at the "Summer School Workshop I, 2024" at Ghent Analysis and PDE Center, carried out during his visit within the framework of "Ghent University's Eureast Platform Short-term Fellowship Programme." The author thanks the conference "Uncertainty and Random Structures: Signal Analysis, Representation Theory and Applications" (2022) for an opportunity to visit Mendeleev's museum in Saint-Petersburg (see [27]).

References

1. The blog of Eurotax. The history of alcohol excise duties in Europe: a long fiscal journey (2024). Retrieved from: eurotax.fr/en/the-history-of-alcohol-excise-duties-in-europe-a-long-fiscal-journeyeurotax.fr
2. Baumhauer, E.-H.: Mémoire sur la Densité, la Dilatation et de la Vapeur d'Alcool et des Mélanges d'Alcool et d'Eau. C.G. Van Der Post, Amsterdam (1860)
3. Blagden, Ch.: Supplementary report on the best method proportioning the excise upon spirituous liquors. Philos. Trans. Royal Soc. **82**, 425–455 (1792) https://doi.org/10.1098/rstl.1792.0022
4. Chatzakou, M., Sarantopoulos, Y.: Bernstein and Markov-type inequalities for polynomials on $L_p(\mu)$ spaces. Dolomites Res. Notes Approx. **12**, 16–28 (2019)
5. Deconinck, K., Poelmans, E., Swinnen, J.: How beer created Belgium (and the Netherlands): The contribution of beer taxes to war finance during the Dutch Revolt. Business Hist. **58**(5), 694–724 (2015). https://doi.org/10.1080/00076791.2015.1024231
6. Gilpin, G.: Tables for reducing the quantities by weight, in any mixture of pure spirit and water, to those by measure; and for determining the proportion, by measure, of each of the two substances in such mixtures. Philos. Trans. Royal Soc. **84**, 275–382 (1794) https://doi.org/10.1098/rstl.1794.0023
7. Mendeleev, D.I.: As a head of the Bureau of Weights and Measures. In S. Knyazev (Ed.), 180th Anniversary of D. I. Mendeleev Institute for Metrology (Events, Personalities, Memoirs). Gumanitarnaya akademiya: Saint-Petersburg (2022). Retrieved from https://www.vniim.ru/book-180vniim.htmlvniim.ru. Based on the book: Ginak, E.: The Metrological Reform of D. I. Mendeleev. Znanie, Moscow (2013)
8. Mendeleev, D.I.: Rassuzhdenie o soedinenii spirta s vodoy (in Russian) [On the Combinations of Water with Alcohol]. Obshchestvennaya Polza, Saint Petersburg (1865)
9. Shiltsev, V., Shiltseva, E.: Mendeleev and vodka: a synergy of science and myth. Sci. First Hand **83**(3) (2019). Retrieved from: https://scfh.ru/papers/mendeleev-i-vodka-sinergiya-nauki-i-mifa/sfch.ru
10. Mendeleev, D.I.: Selected Works, volume 3 (Issledovanie vodnyh rastvorov po udelnomu vesu) (in Russian) [A research on aqueous solutions by their specific gravity]. ONTI-Goskhimtehizdat, Leningrad (1934)
11. Markov, A.A.: Ob odnom voprose D. I. Mendeleeva (in Russian) [On a problem of D. I. Mendeleev]. Acad. Sci. St. Petersburg **62**, 1–24 (1889)
12. Needham, P.: Hydrogen bonding: Homing in on a tricky chemical concept. Stud. History Philos. Sci. Part A **44**(1), 51–65 (2013)
13. Finneran, I., Carroll, B., Allodi, M., Blake, G.: Hydrogen bonding in the ethanol-water dimer. Phys. Chem. Chem. Phys. **17**, 24210–24214 (2015)
14. Gereben, O., Pusztai, L.: Hydrogen bond connectivities in water-ethanol mixtures: On the influence of the H-bond definition. J. Molecular Liq. **220**, 836–841 (2016)
15. Dixit, S., Crain, J., Poon, W., et al.: Molecular segregation observed in a concentrated alcohol-water solution. Nature **416**, 829–832 (2002)
16. Boas, R.P.: Inequalities for the derivatives of polynomials. Math. Mag. **42**(4), 165–174 (1969). https://doi.org/10.2307/2688534
17. Saffari, B.: Some polynomial extremal problems which emerged in the twentieth century. In J.S. Byrnes (Ed.), Twentieth Century Harmonic Analysis–A Celebration. NATO Science Series, **33**. Springer, Dordrecht (2001)
18. Shadrin, A.: Twelve Proofs of the Markov Inequality. In D. K. Dimitrov, G. Nikolov, and R. Uluchev (Eds.), Approximation Theory: A Volume Dedicated to Borislav Bojanov, (pp. 58–63). Marin Drinov Academic Publ. House, Sofia (2004)
19. Markov, V.: Über Polynome, die in einen gegebenen Intervalle möglichst wenig von Null abweichen. Math. Ann. **77**, 213–258 (1916)
20. Erdös, P.: On extremal properties of the derivatives of polynomials. Ann. of Math. **41**(2), 310–313 (1940)

21. Borwein, P.: Markov's inequality for polynomials with real zeros Proc. Amer. Math. Soc. **93**, 43–47 (1985)
22. Konyagin, S.V.: Markov's inequality for polynomials in the metric of L. Proc. Steklov Inst. Math. **145**, 129–138 (1981)
23. Aleksov, D., Nikolov, G., Shadrin, A.: On the Markov inequality in the L2-norm with the Gegenbauer weight. J. Approx. Theory **208**, 9–20 (2016)
24. Wales, W.: George Gilpin. Cook's Log **43**(4), 32–35 (2020). Published by: Captain Cook Study Unit, Ipswich, ISSN 1358-0639
25. Wilhelmsen, D.: A Markov inequality in several dimensions. J. Approx. Theory **11**(3), 216–220 (1974)
26. Parke, S., Birch, G.: Solution properties of ethanol in water. Food Chem. **67**(3), 241–246 (1999). https://doi.org/10.1016/S0308-8146(99)00124-7
27. Vagharshakyan, A.: Uniqueness theorems in analysis. A talk at the international conference Uncertainty and random structures: Signal Analysis, Representation Theory and Applications. Saint-Petersburg State University, Saint-Petersburg, December 12-16 (2022)

Geometric Estimates on a Nonlinear Parabolic Equation Under Perelman-Ricci Super Flow with Applications to Global and Ancient Solutions

Vahideh Vahidifar

Abstract In this extended abstract, we present recent results on local and global gradient estimates of different geometric types for positive solutions to a nonlinear parabolic equation on a smooth metric measure space whose underlying metric and potential evolve in time. We present some applications, including parabolic Harnack inequalities and Liouville-type theorems and discuss some related examples.

Keywords Smooth metric measure space · f-Laplacian · Perelman-Ricci super flow · Bakry-Émery tensor · Souplet-Zhang estimates · Li-Yau estimates · Liouville-type results

MSC2020 53C44 · 58J60 · 58J35 · 60J60

1 Introduction

Let (M, g) be a complete Riemannian manifold of dimension $n \geq 2$, $d\mu = e^{-f} dv_g$ be a weighted measure associated with the metric tensor g and potential function f and let dv_g be the standard Riemannain volume measure. The triple $(M, g, d\mu)$ is referred to as a smooth metric space (or a weighted manifold). In this paper, we present local and global gradient estimates of Hamilton-Souplet-Zhang and Li-Yau types for positive solutions $w = w(x, t)$ to the nonlinear parabolic equation

$$\left[\frac{\partial}{\partial t} - q(x,t) - \Delta_f\right] w(x,t) = \Sigma(t, x, w(x,t)), \qquad x \in M, \quad t \in \mathbb{R}. \quad (1)$$

Here Δ_f denotes the f-Laplacian (also known as the weighted or Witten Laplacian or Nelson's diffusion operator in stochastic mechanics). The f-Laplacian acts on functions $w \in \mathscr{C}^2(M)$ through the identity

V. Vahidifar (✉)
School of Mathematical and Physical Sciences, University of Sussex, Falmer, United Kingdom
e-mail: v.vahidifar@sussex.ac.uk

M. Chatzakou et al. (eds.), *Direct and Inverse Problems with Applications*,
Trends in Mathematics 12, https://doi.org/10.1007/978-3-031-98645-1_23

$$\Delta_f w = \mathrm{div}_f \nabla w = e^f \mathrm{div}(e^{-f} \nabla w) = \Delta w - \langle \nabla f, \nabla w \rangle,$$

where Δ, div and ∇ are the usual Laplace-Beltrami, divergence and gradient operators associated with the metric g respectively.

The f-Laplacian Δ_f is a symmetric diffusion operator with respect to the invariant weighted measure $d\mu$ and arises in a variety of contexts ranging from probability theory, geometry and stochastic processes to quantum field theory and statistical mechanics. It is a natural generalisation of the Laplacian to the smooth metric measure space setting and coincides with the latter precisely when the potential f is a constant function. It can be easily seen that for smooth compactly supported $u, w \in \mathscr{C}_0^\infty(M)$:

$$\int_M e^{-f} w \Delta_f u \, dv_g = -\int_M e^{-f} \langle \nabla u, \nabla w \rangle \, dv_g = \int_M e^{-f} u \Delta_f w \, dv_g.$$

As for the geometry and curvature properties of the triple $(M, g, d\mu)$, for fixed constant $m \geq n \geq 2$ (not necessarily an integer), we define the second order symmetric tensor field $\mathscr{R}ic_f^m(g)$, called the Bakry-Émery m-Ricci curvature tensor, by

$$\mathscr{R}ic_f^m(g) = \mathscr{R}ic(g) + \nabla\nabla f - \frac{\nabla f \otimes \nabla f}{m - n}. \tag{2}$$

Here $\mathscr{R}ic(g)$ is the usual Ricci tensor of g and $\nabla\nabla f$ is the Hessian of f. To clear any ambiguity, in (2) when $m = n$, by convention, f can only be a constant, thus making $\mathscr{R}ic_f^m(g) = \mathscr{R}ic(g)$. We also set, as the counterpart of (2) for $m = \infty$,

$$\mathscr{R}ic_f(g) = \mathscr{R}ic(g) + \nabla\nabla f.$$

The weighted Bochner-Weitzenböck formula relating Δ_f and $\mathscr{R}ic_f(g)$, asserts that for every $w \in \mathscr{C}^3(M)$ we have:

$$\frac{1}{2}\Delta_f |\nabla w|^2 - \langle \nabla w, \nabla \Delta_f w \rangle = |\nabla\nabla w|^2 + \mathscr{R}ic_f(\nabla w, \nabla w).$$

There is no counterpart for this identity to relate Δ_f and $\mathscr{R}ic_f^m(g)$ when $m < \infty$ but one can easily prove the *inequality*:

$$\frac{1}{2}\Delta_f |\nabla w|^2 - \langle \nabla w, \nabla \Delta_f w \rangle \geq \frac{1}{m}(\Delta_f w)^2 + \mathscr{R}ic_f^m(\nabla w, \nabla w).$$

Perelman-Ricci Flows and Super Flows. A (k, m) Perelman-Ricci super flow refers to the evolution of a one-parameter family of smooth metric tensors $g = g(t)$ and potential functions $f = f(t)$ on M subject to the inequality

$$\frac{1}{2}\partial_t g(x, t) + \mathscr{R}ic_f^m(g)(x, t) \geq -kg(x, t). \tag{3}$$

Here k is a real constant with $m \geq n$ (not necessarily an integer). We interpret $m = \infty$ by replacing $\mathscr{R}ic_f^m(g)$ with $\mathscr{R}ic_f(g)$. Upon switching the inequality in (3) to equality, we get the (k, m) Perelman-Ricci flow [(4) below]. Under the additional assumption that the weighted measure $d\mu = e^{-f} dv_g$ remains static in time, it can be seen that

$$\partial_t(d\mu_t) = \partial_t\left(e^{-f(t)} dv\right) = \left[-\partial_t f + \frac{1}{2}\mathrm{Tr}\,(\partial_t g)\right] d\mu_t = 0.$$

Here Tr stands for the metric trace on $(0, 2)$-tensors and the combined system resulting from the flow and the stationary volume measure constraint takes the form

$$\frac{1}{2}\partial_t g(x,t) + \mathscr{R}ic_f^m(g)(x,t) = -kg(x,t), \tag{4}$$

$$\partial_t f(x,t) - \frac{1}{2}\mathrm{Tr}\,(\partial_t g)\,(x,t) = 0. \tag{5}$$

In particular it follows from (4)–(5) that the potential f satisfies the evolution equation

$$\partial_t f + \Delta f = \mathrm{Tr}\left(\frac{1}{2}\partial_t g + \nabla\nabla f\right) = -R + \frac{|\nabla f|^2}{m-n} - nk.$$

The case $(k, m) = (0, \infty)$ in the system (4)–(5) is what was introduced by Perelman in [5] as the L^2-gradient flow of the functional $\mathscr{P}(g, f) = \int_M (R + |\nabla f|^2)e^{-f} dv_g$ subject to the stationary volume measure constraint considered above. Here $R = R(g)$ is the scalar curvature, $\mathscr{R}ic_f(g) = \mathscr{R}ic(g) + \nabla\nabla f$, and f is the potential satisfying the so-called conjugate or adjoint heat equation,

$$\Box^\star f = -\partial_t f - \Delta f + R = 0.$$

The above discussion illustrates another aspect of how the problems considered here link with the larger and more recent body of research on the subject and the remarkable work on the Poincaré conjecture [5]. For further reading and related work, see [2, 18, 19] as well as [9, 10, 13, 16, 17] and the references therein.

Notation. Let us introduce and fix some notation and terminology that will be used. For $X \in \mathbb{R}$ we write $X_+ = \max(X, 0)$ and $X_- = \min(X, 0)$ and thus $X_+ + X_- = X$. We denote by $d = d(x_1, x_2, t)$ the geodesic distance between x_1, x_2 on M with respect to $g = g(t)$. Fixing x_0 on M we write $r = r(x, x_0, t)$ for the geodesic radial variable at x with origin x_0 at time t. For $x_0 \in M$, $t_0 \in \mathbb{R}$ and $R > 0$, $T > 0$ we define the space-time cylinder $Q_{R,T} \equiv \{(x,t) : d(x, x_0, t) \leq R$ and $t_0 - T \leq t \leq t_0\} \subset M \times (-\infty, \infty)$. When g is time independent we denote by $\mathscr{B}_r(x_0) \subset M$ the geodesic ball of radius $r > 0$ centred at x_0. In this case $Q_{R,T} = \mathscr{B}_R(x_0) \times [t_0 - T, t_0]$. When the choice of x_0 is clear from the context we often abbreviate and write $d(x,t)$, $r(x,t)$ or $\mathscr{B}_r$ for convenience.

2 Statement of the Results

In this section, we present the main results. We start by formulating a local elliptic gradient estimate of Hamilton type for positive solutions to the equation (1) where the metric g and potential f are time-dependent and of class $\mathscr{C}^2$ [*see* Remark 1]. The detailed proofs of all results in this section can be found in our papers [11–16].

2.1 A Local and Global Souplet-Zhang Type Gradient Estimate for (1)

For the purpose of this estimate, we fix $x_0 \in M$, $t_0 \in \mathbb{R}$ and restrict attention to the compact set $Q_{R,T} \equiv \{(x,t) : d(x, x_0, t) \le R \text{ and } t_0 - T \le t \le t_0\}$. Here $R \ge 2$, $T > 0$. The estimate makes use of the constants $\mathsf{k}, \mathsf{h} \ge 0$ used to describe the lower bounds (in the sense of symmetric tensors):

$$\mathscr{R}ic_f(g)(x,t) \ge -(n-1)\mathsf{k}g(x,t), \qquad \partial_t g(x,t) \ge -2\mathsf{h}g(x,t),$$

for $(x,t) \in Q_{R,T}$. Note that due to the local nature of the estimate here all that is required of w is to be a positive smooth solution to (1) in an open set containing $Q_{R,T}$.

Theorem 1 *Let* $(M, g, d\mu)$ *with* $d\mu = e^{-f} dv_g$ *be a complete smooth metric measure space satisfying* $\mathscr{R}ic_f(g)(x,t) \ge -(n-1)\mathsf{k}g(x,t)$ *and* $\partial_t g(x,t) \ge -2\mathsf{h}g(x,t)$ *in* $Q_{R,T}$ *for some* $\mathsf{k}, \mathsf{h} \ge 0$. *Let* w *be a positive and bounded solution to* (1) *with* $0 < w \le D$. *Then for all* (x,t) *in* $Q_{R/2,T}$ *with* $t > t_0 - T$ *we have:*

$$\frac{|\nabla w|}{w}(x,t) \le C\left(1 - \log\frac{w}{D}(x,t)\right) \times \left\{ \begin{array}{l} \dfrac{1}{R} + \sqrt{\dfrac{[\gamma_{\Delta_f}]_+}{R}} + \sqrt{\mathsf{h}+\mathsf{k}} \\ + \dfrac{1}{\sqrt{t - t_0 + T}} + \sup_{Q_{R,T}} \dfrac{[q(x,t)]_+^{1/2}}{[1 - \log(w/D)]} \\ + \sup_{Q_{R,T}} \dfrac{|\nabla q(x,t)|^{1/3}}{[1 - \log(w/D)]^{2/3}} + \sup_{Q_{R,T}} \left[\dfrac{|\Sigma_x(t,x,w)|}{w[1 - \log(w/D)]^2}\right]^{\frac{1}{3}} \\ + \sup_{Q_{R,T}} \left[\dfrac{w[1 - \log(w/D)]\Sigma_w(t,x,w) + \log(w/D)\Sigma(t,x,w)}{w[1 - \log(w/D)]^2}\right]_+^{\frac{1}{2}} \end{array} \right\}.$$

Here $C = C(n) > 0$ *is a constant depending only on* n *and* $[\gamma_{\Delta_f}]_+ = \max(\gamma_{\Delta_f}, 0)$ *where the functional quantity* γ_{Δ_f} *is defined by*

$$\gamma_{\Delta_f} = \max_{(x,t)}\{\Delta_f r(x,t) : d(x,t) = 1, t_0 - T \le t \le t_0\}.$$

Subject to the bounds in Theorem 1 being global in space by passing to the limit $R \to \infty$ we have the following global (in space) estimate.

Theorem 2 *Let $(M, g, d\mu)$ with $d\mu = e^{-f} dv_g$ be a complete smooth metric measure space satisfying $\mathscr{R}ic_f(g)(x,t) \geq -(n-1)\mathsf{k}g(x,t)$ and $\partial_t g(x,t) \geq -2\mathsf{h}g(x,t)$ in $M \times [t_0 - T, t_0]$ for some $\mathsf{k}, \mathsf{h} \geq 0$. Let w be a positive and bounded solution to* (1) *with $0 < w \leq D$, then for every $x \in M$ and $t_0 - T < t \leq t_0$ we have:*

$$\frac{|\nabla w|}{w}(x,t) \leq C\left(1 - \log\frac{w}{D}(x,t)\right) \times$$
$$\left\{\begin{array}{l} \sqrt{\mathsf{h}+\mathsf{k}} + \dfrac{1}{\sqrt{t - t_0 + T}} + \sup_{M\times[t_0-T,t_0]} \dfrac{[q(x,t)]_+^{1/2}}{[1-\log(w/D)]} \\ + \sup_{M\times[t_0-T,t_0]} \dfrac{|\nabla q(x,t)|^{1/3}}{[1-\log(w/D)]^{2/3}} + \sup_{M\times[t_0-T,t_0]} \left[\dfrac{|\Sigma_x(t,x,w)|}{w[1-\log(w/D)]^2}\right]^{\frac{1}{3}} \\ + \sup_{M\times[t_0-T,t_0]} \left[\dfrac{w[1-\log(w/D)]\Sigma_w(t,x,w) + \log(w/D)\Sigma(t,x,w)}{w[1-\log(w/D)]^2}\right]_+^{\frac{1}{2}} \end{array}\right\}.$$

Remark 1 From the bounds $\mathscr{R}ic_f(g) \geq -(n-1)\mathsf{k}g$ and $\partial_t g \geq -2\mathsf{h}g$, we obtain the Perelman-Ricci super flow inequality (in the sense of symmetric tensors):

$$\partial_t g(x,t) + 2\mathscr{R}ic_f(g)(x,t) \geq -2kg(x,t),$$

with $k = (n-1)\mathsf{k} + \mathsf{h}$. The lower bound $\mathscr{R}ic_f(g) \geq -(n-1)\mathsf{k}g$ here is required for the Wei-Wylie weighted Laplacian comparison theorem and the bound $\partial_t g \geq -2\mathsf{h}g$ for controlling the time derivative of the geodesic distance, namely, $d_t = \partial_t[d(x, x_0, t)]$.

Remark 2 In the static case $\partial_t g \equiv 0$ and $\partial_t f \equiv 0$ we can set $\mathsf{h} = 0$. Moreover when $\mathscr{R}ic_f(g)(x,t) \geq 0$ in $Q_{R,T}$ we can take $\mathsf{k} = 0$.

2.2 *A Local and Global Li-Yau Estimate for* (1)

We now move on to the Li-Yau estimate (also known as a differential Harnack inequality). To this end we pick $x_0 \in M$ and set $Q_{2R,T} = Q_{2R,T}(x_0)$ where $R > 0$, $T > 0$. The estimate makes use of the lower bound on $\mathscr{R}ic_f^m(g)$ (with $n \leq m < \infty$), along with the bounds on w, $\partial_t g$, ∇f, $\nabla \partial_t g$ and $\nabla \partial_t f$ all in $Q_{2R,T}$. [See (12)–(13).]

To proceed, let us introduce for given $m \geq n$, $\lambda > 1$ and non-negative constants $k_1, \overline{k}_2, \underline{k}_2, k_3, \ell_1, \ell_2$ the numerical quantities

$$\mathsf{A} = 2[(m-1)k_1 + (\lambda - 1)\overline{k}_2 + k_3], \tag{6}$$
$$\mathsf{B} = \lambda\ell_2/2 + \lambda\underline{k}_2\ell_1,$$
$$\mathsf{C} = \lambda^2 n(\underline{k}_2 + \overline{k}_2)^2 + 2\lambda^2 n k_3. \tag{7}$$

Moreover, for given $\Sigma = \Sigma(t, x, w)$, let us further introduce the quantities

$$\gamma_1^\Sigma = -\inf_{\Theta_{2R,T}} \left\{ \frac{1}{w} [\Sigma(t, x, w) - w\Sigma_w(t, x, w) + \lambda w^2 \Sigma_{ww}(t, x, w)]_- \right\}, \tag{8}$$

$$\gamma_2^\Sigma = \sup_{\Theta_{2R,T}} \left\{ \frac{1}{w} |\Sigma_x(t, x, w) - \lambda w \Sigma_{xw}(t, x, w)| \right\},$$

$$\gamma_3^\Sigma = \sup_{\Theta_{2R,T}} \left\{ \frac{1}{w} [w\Sigma_w(t, x, w) - \Sigma(t, x, w)]_+ \right\},$$

$$\gamma_4^\Sigma = -\inf_{\Theta_{2R,T}} \left[\frac{1}{w} \Delta_f \Sigma^x(t, x, w) \right]_-, \tag{9}$$

where $\Theta_{2R,T} = \{(t, x, w) : (x, t) \in Q_{2R,T},\ \underline{w} \le w \le \overline{w}\} \subset [0, T] \times M \times (0, \infty)$. Here $\overline{w}$, $\underline{w}$ denote the maximum and minimum of w on the compact space-time cylinder $Q_{2R,T}$. Note that in the liner case $\Sigma(t, x, w) = q(x, t)w$ we have $\gamma_1^{qw} = 0$, $\gamma_3^{qw} = 0$ and

$$\gamma_2^{qw} = (\lambda - 1) \sup_{\Theta_{2R,T}} |\nabla q|, \tag{10}$$

$$\gamma_4^{qw} = -\inf_{\Theta_{2R,T}} [\Delta_f q]_-. \tag{11}$$

Theorem 3 *Let $(M, g, d\mu)$ be a complete smooth metric measure space with $d\mu = e^{-f} dv_g$ and suppose that the metric g and potential f are time dependent and of class $\mathscr{C}^2$. Assume that $\mathscr{R}ic_f^m(g) \ge -(m-1)k_1 g$ and that*

$$-2\underline{k}_2 g \le \partial_t g \le 2\overline{k}_2 g, \qquad |\nabla \partial_t g| \le 2k_3, \tag{12}$$

$$|\nabla f| \le \ell_1, \qquad |\nabla \partial_t f| \le \ell_2, \tag{13}$$

in $Q_{2R,T}$ for suitable constants $k_1, \underline{k}_2, \overline{k}_2, k_3 \ge 0$, $\ell_1, \ell_2 \ge 0$. Let $w = w(x, t)$ be a positive solution to (1) in $Q_{2R,T}$. Then for every $\lambda > 1$, $\varepsilon \in (0, 1)$ and all (x, t) in $Q_{R,T}$ with $t > 0$ we have

$$\begin{aligned}
\frac{|\nabla w|^2}{\lambda w^2} - \frac{\partial_t w}{w} + q + \frac{\Sigma}{w} &\le m\lambda \left[\frac{1}{t} + c_1 \underline{k}_2 + \gamma_3^\Sigma \right] \\
&+ \frac{m\lambda}{R^2} \left[\frac{m c_1^2 \lambda^2}{2(\lambda - 1)} + c_2 + (m-1)c_1(1 + R\sqrt{k_1}) + 2c_1^2 \right] \\
&+ \sqrt{m} \left\{ \frac{m\lambda^2 (\mathsf{A} + \gamma_1^\Sigma)^2}{4(1-\varepsilon)(\lambda-1)^2} + \frac{3}{4} \left[\frac{4m\lambda^2 (\mathsf{B} + \gamma_2^\Sigma + \gamma_2^{qw})^4}{\varepsilon(\lambda - 1)^2} \right]^{1/3} + \mathsf{C} + \lambda(\gamma_4^\Sigma + \gamma_4^{qw}) \right\}^{1/2}.
\end{aligned}$$

Note that unlike Theorem 1, here, the lower bound is imposed on the Bakry-Émery m-Ricci tensor $\mathscr{R}ic^m_f(g)$ with $n \le m < \infty$ and not on $\mathscr{R}ic_f(g)$. This is evidently a stronger assumption.

The global counterpart of the above local estimate can be obtained by imposing appropriate global bounds in the assumptions as is formulated in the following theorem.

Theorem 4 *Let $(M, g, d\mu)$ be a complete smooth metric measure space with $d\mu = e^{-f} dv_g$ and assume that the metric g and potential f are time-dependent and of class $\mathscr{C}^2$. Assume $\mathscr{R}ic^m_f(g) \ge -(m-1)k_1 g$, (12) and (13) hold globally in $M \times [0, T]$. Let $w = w(x, t)$ be a positive solution to (1). Then for every $\lambda > 1$, $\varepsilon \in (0, 1)$ and all $x \in M$, $0 < t \le T$ we have*

$$\frac{|\nabla w|^2}{\lambda w^2} - \frac{\partial_t w}{w} + q + \frac{\Sigma}{w} \le m\lambda \left[\frac{1}{t} + c_1 \underline{k}_2 + \gamma_3^\Sigma\right]$$
$$+ \sqrt{m}\left\{\frac{m\lambda^2(\mathsf{A} + \gamma_1^\Sigma)^2}{4(1-\varepsilon)(\lambda-1)^2} + \frac{3}{4}\left[\frac{4m\lambda^2(\mathsf{B} + \gamma_2^\Sigma + \gamma_2^{qw})^4}{\varepsilon(\lambda-1)^2}\right]^{1/3} + \mathsf{C} + \lambda(\gamma_4^\Sigma + \gamma_4^{qw})\right\}^{1/2}.$$

Here A, B, C are as in (6)–(7), γ_1^Σ, γ_2^Σ, γ_3^Σ, γ_4^Σ as in (8)–(9), and γ_2^{qw}, γ_4^{qw} are as in (10)–(11) with the supremum and infimums taken over $M \times [0, T]$ respectively.

Theorem 5 *Let w be a positive solution to (1). Then for (x_1, t_1), (x_2, t_2) in $Q_{R,T}$ with $t_2 > t_1$ and $\lambda > 1$ we have*

$$\frac{w(x_2, t_2)}{w(x_1, t_1)} \ge \left(\frac{t_2}{t_1}\right)^{-m\lambda} e^{-\lambda L(x_1, x_2, t_2 - t_1)} e^{(t_2 - t_1)\mathsf{S}}.$$

Here S is given by

$$\mathsf{S} = -\frac{m\lambda}{R^2}\left[\frac{mc_1^2\lambda^2}{2(\lambda-1)} + c_2 + (m-1)c_1(1 + R\sqrt{k_1}) + 2c_1^2\right] - m\lambda(\gamma_3^\Sigma + c_1\underline{k}_2) + \gamma_5^\Sigma$$
$$+ \gamma_5^{qw} - \sqrt{m}\left\{\frac{m\lambda^2(\mathsf{A} + \gamma_1^\Sigma)^2}{4(1-\varepsilon)(\lambda-1)^2} + \frac{3}{4}\left[\frac{4m\lambda^2(\mathsf{B} + \gamma_2^\Sigma + \gamma_2^{qw})^4}{\varepsilon(\lambda-1)^2}\right]^{1/3} + \mathsf{C} + \lambda(\gamma_4^\Sigma + \gamma_4^{qw})\right\}^{1/2},$$

where

$$\gamma_5^\Sigma = \inf_{\Theta_{2R,T}} \left\{\frac{\Sigma(t, x, w)}{w}\right\}, \qquad \gamma_5^{qw} = \inf_{\Theta_{2R,T}} q.$$

Furthermore,

$$L(x_1, x_2, t_2 - t_1) = \inf_{\gamma \in \Gamma}\left[\frac{1}{4(t_2 - t_1)}\int_0^1 |\dot{\gamma}|^2\right],$$

where Γ is the set of all $\gamma \in \mathscr{C}^1([0,1]; M)$ lying entirely in $Q_{R,T}$ with $\gamma(0) = x_1$ and $\gamma(1) = x_2$. If the bounds in Theorem 4 are global then the above estimate is global.

2.3 Applications to Liouville-Type Results (I): $\mathscr{R}ic_f(g) \geq 0$

Let us now move on to discussing some applications of the gradient estimates above. Here, we mainly focus on applications to Liouville and global constancy type results for both the parabolic and elliptic (stationary states) equations associated with (1). To this end, assume w is a smooth bounded positive solution to the elliptic equation

$$\Delta_f w + \Sigma(w) = \Delta w - \langle \nabla f, \nabla w \rangle + \Sigma(w) = 0. \tag{14}$$

Note that since a constant solution to (14) must be a zero of Σ, if Σ has no zeros $w > 0$ then a Liouville theorem can be seen as a non-existence result.

Theorem 6 *Let $(M, g, d\mu)$ with $d\mu = e^{-f} dv_g$ be a complete smooth metric measure space satisfying $\mathscr{R}ic_f(g) \geq 0$ in M. Let w be a positive and bounded solution to (14) with $0 < w \leq D$ and assume $[1 - \log(w/D)]w\Sigma'(w) + \log(w/D)\Sigma(w) \leq 0$. Then w is a constant. In particular $\Sigma(w) = 0$.*

Let us now discuss the implications and some examples of the above theorems to specific nonlinearities of interest arising in various contexts within geometry and mathematical physics.

Theorem 7 *Let $(M, g, d\mu)$ with $d\mu = e^{-f} dv_g$ be a complete smooth metric measure space satisfying $\mathscr{R}ic_f(g) \geq 0$ in M. Let w be a positive and bounded solution to the equation*

$$\Delta_f w + \sum_{j=1}^{N} \mathsf{A}_j w^{\mathsf{p}_j} + \sum_{j=1}^{N} \mathsf{B}_j w^{\mathsf{q}_j} = 0.$$

Then if $\mathsf{A}_j \geq 0$, $\mathsf{B}_j \leq 0$, $\mathsf{p}_j \leq 0$ and $\mathsf{q}_j \geq 1$ then w is a constant.

Next, we focus on nonlinearities in the "*split*" form $\Sigma(w) = X(w) + w^r Y(\log w)$ where $X = X(w) \in \mathscr{C}^1(0, \infty)$, r is a real exponent and $Y = Y(s) = Y(\log w) \in \mathscr{C}^1(-\infty, \infty)$. The next result is an application of Theorem 6.

Theorem 8 *Let $(M, g, d\mu)$ with $d\mu = e^{-f} dv_g$ be a complete smooth metric measure space satisfying $\mathscr{R}ic_f(g) \geq 0$ in M. Let w be a positive bounded solution to the nonlinear elliptic equation*

$$\Delta_f w + X(w) + w^r Y(\log w) = 0.$$

Suppose $X'(w) \leq 0$ *and* $X(w) - wX'(w) \geq 0$. *Moreover, when* $Y \not\equiv 0$, *depending on* w *satisfying* $w \geq 1$ *or* $0 < w \leq 1$ *everywhere on* M, *assume that:*

$$(\mathrm{Y1}) := \begin{cases} r \geq 1,\ w \geq 1, \\ Y(s) \leq 0 \text{ and} \\ Y'(s) \leq 0 \text{ for all } s \geq 0, \end{cases} \qquad \mathrm{Y2}) := \begin{cases} r \leq \min(\gamma, 1),\ 0 < w \leq 1, \\ Y(s) \geq 0,\ Y'(s) \leq 0 \text{ and} \\ sY'(s) \geq \gamma Y(s), \text{ all } s \leq 0, \text{ some } \gamma \geq 0. \end{cases}$$

Then w *is a constant. In particular* $X(w) + w^r Y(\log w) = 0$.

Theorem 9 *Let* $(M, g, d\mu)$ *with* $d\mu = e^{-f} dv_g$ *be a complete smooth metric measure space satisfying* $\mathscr{R}ic_f(g) \geq 0$ *in* M. *Let* $\Sigma'(w) \leq 0$, $\Sigma(w) - w\Sigma'(w) \geq 0$ *and* $\Sigma(w) \geq a$ *for some* $a > 0$ *and all* $w > 0$. *Then the equation*

$$\partial_t w - \Delta_f w = \Sigma(w),$$

does not have any positive ancient solutions with the growth $w(x,t) = e^{o(\sqrt{r(x)}+\sqrt{|t|})}$.

2.4 Applications to Liouville-Type Results (II): $\mathscr{R}ic^m_f(g) \geq 0$

We complement the Liouville-type results given earlier by invoking the Li-Yau gradient estimates above which were obtained under a different curvature condition.

Theorem 10 *Let* $(M, g, d\mu)$ *with* $d\mu = e^{-f} dv_g$ *be a complete smooth metric measure space satisfying* $\mathscr{R}ic^m_f(g) \geq -(m-1)kg$ *in* M. *Let* w *be a smooth positive solution to* (14). *Then, for every* $\alpha > 1$ *and* $\varepsilon \in (0,1)$ *the following global estimate holds on* M:

$$\frac{|\nabla w|^2}{\lambda w^2} + \frac{\Sigma(w)}{w} \leq m\lambda \left[\frac{2(m-1)k + \gamma_1^\Sigma}{2(\lambda - 1)\sqrt{1-\varepsilon}} + \gamma_3^\Sigma \right].$$

Theorem 11 *Let* $(M, g, d\mu)$ *with* $d\mu = e^{-f} dv_g$ *be a complete smooth metric measure space satisfying* $\mathscr{R}ic^m_f(g) \geq 0$ *in* M. *Let* w *be a smooth positive solution to* (14) *and assume that along the solution* w *we have* $\Sigma(w) \geq 0$, $\Sigma(w) - w\Sigma_w(w) \geq 0$ *and*

$$\Sigma(w) - w\Sigma_w(w) + \alpha w^2 \Sigma_{ww}(w) \geq 0,$$

for some $\alpha > 1$, *everywhere on* M. *Then* w *must be a constant and* $\Sigma(w) = 0$.

Theorem 11 leads to the following conclusion on Yamabe equations. As indicated earlier, since a constant solution to (14) must be a zero of Σ, when Σ has no zeros $w > 0$, a Liouville theorem can be seen as a non-existence result for positive solutions. In the following theorem this happens precisely when $\mathsf{A}_j > 0$ for at least one $1 \leq j \leq N$.

Theorem 12 *Let $(M, g, d\mu)$ with $d\mu = e^{-f} dv_g$ be a complete smooth metric measure space satisfying $\mathscr{R}ic_f^m(g) \geq 0$ in M. Let w be a positive smooth solution to the equation*

$$\Delta_f w + \sum_{j=1}^{N} \mathsf{A}_j w^{p_j} = 0.$$

If $\mathsf{A}_j \geq 0$ and $p_j \leq 1$ for all $1 \leq j \leq N$ then w must be a constant and $\sum \mathsf{A}_j w^{p_j} = 0$.

As another application, consider a nonlinearity in the form of a superposition of a logarithmic and a power-like nonlinearity with constant coefficients A, B, exponent s and $Y \in \mathscr{C}^2(\mathbb{R})$ in the form

$$\Sigma(w) = \mathsf{A}wY(\log w) + \mathsf{B}w^s.$$

The following theorem now directly results from Theorem 11.

Theorem 13 *Let $(M, g, d\mu)$ with $d\mu = e^{-f} dv_g$ be a complete smooth metric measure space satisfying $\mathscr{R}ic_f^m(g) \geq 0$. Let w be a positive smooth solution to the equation*

$$\Delta_f w + \mathsf{A}wY(\log w) + \mathsf{B}w^s = 0.$$

Assume that along the solution w we have $Y \geq 0$, $Y' \leq 0$ and $\alpha Y'' + (\alpha - 1)Y' \geq 0$ for some $\alpha > 1$. Furthermore assume $\mathsf{A}, \mathsf{B} \geq 0$ and $s \leq 1$. Then w must be a constant and $\mathsf{A}wY(\log w) + \mathsf{B}w^s = 0$.

Acknowledgements The author wishes to express her immense gratitude to Professor Michael Ruzhansky for the opportunity to visit the Ghent Analysis and PDE Center for three months during the summer of 2024. She also wishes to thank the organisers of the workshop "Direct and Inverse Problems and Application" for organising such a nice and stimulating meeting and the opportunity to give an invited talk. Support from the Engineering and Physical Sciences Research Council (EPSRC) through the grant EP/V027115/1 is gratefully acknowledged.

References

1. D. Bakry, I., Gentil, M.L.: Analysis and geometry of Markov diffusion operators, A Series of comprehensive studies in mathematics, **348**, Springer (2012)
2. Chow, B., Lu, P., Nei, L.: Hamilton's Ricci flow, Graduate Studies in Mathematics **77**, AMS (2006)
3. Hamilton, R.: A matrix Harnack estimate for heat equation. Comm. Anal. Geom. 113–126 (1993)
4. Li, P., Yau, S.T.: On the parabolic kernel of Schrödinger operator. Acta Math. **156**, 153–201 (1986)
5. Perelman, G.: The entropy formula for the Ricci Flow and its geometric application, arXiv: math. DG/0211159v1 (2002)

6. Souplet, P., Zhang, Q.S.: Sharp gradient estimate and Yau's Liouville theorem for the heat equation on noncompact manifolds. Bull. Lond. Math. Soc. **38**, 1045–1053 (2006)
7. Taheri, A.: Function spaces and partial differential equations, Vol. **I**, Oxford Lecture Series in Mathematics and its Applications, **40**, OUP (2015)
8. Taheri, A.: Function spaces and partial differential equations, Vol. **II**, Oxford Lecture Series in Mathematics and its Applications, **41**, OUP (2015)
9. Taheri, A.: Liouville theorems and elliptic gradient estimates for a nonlinear parabolic equation involving the Witten Laplacian, Published online in: Adv. Calc. Var., De Gruyter (2021)
10. Taheri, A.: Gradient estimates for a weighted Γ-nonlinear parabolic equation coupled with a super Perelman-Ricci flow and implications, Published online in: Potential Anal., Springer (2021)
11. Taheri, A., Vahidifar, V.: Gradient estimates for nonlinear elliptic equations involving the Witten Laplacian on smooth metric measure spaces and implications. Adv. Nonlinear Anal, 12(1) (2023)
12. Taheri, A., Vahidifar, V.: Souplet-Zhang and Hamilton-type gradient estimates for non-linear elliptic equations on smooth metric measure spaces. Mathematika **69**(3), 751–779 (2023)
13. Taheri, A., Vahidifar, V.: Gradient estimates for a nonlinear parabolic equation on smooth metric measure spaces with evolving metrics and potentials. Nonlinear Anal. 232 (2023)
14. Taheri, A., Vahidifar, V.: Curvature conditions, Liouville-type theorems and Harnack inequalities for a nonlinear parabolic equation on smooth metric measure spaces. Adv. Nonlinear Studies **24**, 553–591 (2024)
15. Taheri, A., Vahidifar, V.: Hamilton and Li-Yau type gradient estimates for a weighted nonlinear parabolic equation under a super Perelman-Ricci flow. Partial Differ. Equ. Appl, 5(1), Springer (2024)
16. Taheri, A., Vahidifar, V.: Differential Harnack estimates for a weighted nonlinear parabolic equation under a super Perelman-Ricci flow and implications. Proc. R. Soc. Edinb. Sect. A: Math. **155**(2), 687–717 (2025)
17. Taheri, A., Vahidifar, V.: The nonlinear fast diffusion equation on smooth metric measure spaces: Hamilton-Souplet-Zhang estimates and a Ricci-Perelman super flow. Calc. Var. Partial Differ. Equ. **64**(3), 1–36 (2025)
18. Wang, F.Y.: Analysis for diffusion processes on riemannian manifolds. Advanced Series on Statistical Science & Probability, Vol. **18**, World Scientific (2013)
19. Zhang, Q.S.: Sobolev inequalities, heat kernels under Ricci flow and the Poincaré conjecture, CRC Press (2011)

MIX
Papier aus verantwortungsvollen Quellen
Paper from responsible sources
FSC® C105338

If you have any concerns about our products,
you can contact us on
ProductSafety@springernature.com

In case Publisher is established outside the EU,
the EU authorized representative is:
Springer Nature Customer Service Center GmbH
Europaplatz 3, 69115 Heidelberg, Germany

Printed by Libri Plureos GmbH
in Hamburg, Germany